Daily Stoic
A Daily Journal:

ON MEDITATION, STOICISM, WISDOM AND
PHILOSOPHY, TO IMPROVE YOUR LIFE

George Tanner

© 2018

COPYRIGHT

Daily Stoic A Daily Journal: On Meditation, Stoicism, Wisdom and Philosophy, to Improve your Life

By George Tanner

Copyright @2018 By George Tanner

All Rights Reserved.

The following eBook is reproduced below with the goal of providing information that is as accurate and as reliable as possible. Regardless, purchasing this eBook can be seen as consent to the fact that both the publisher and the author of this book are in no way experts on the topics discussed within, and that any recommendations or suggestions made herein are for entertainment purposes only. Professionals should be consulted as needed before undertaking any of the action endorsed herein.

This declaration is deemed fair and valid by both the American Bar Association and the Committee of Publishers Association and is legally binding throughout the United States.

Furthermore, the transmission, duplication or reproduction of any of the following work, including precise information, will be considered an illegal act, irrespective whether it is done electronically or in print. The legality extends to creating a secondary or tertiary copy of the work or a recorded copy and is only allowed with express written consent of the Publisher. All additional rights are reserved.

The information in the following pages is broadly considered to be a truthful and accurate account of facts, and as such any inattention, use or misuse of the information in question by the reader will render any resulting actions solely under their purview. There are no scenarios in which the publisher or the original author of this work can be in any fashion deemed liable for any hardship or damages that may befall them after undertaking information described herein.

Additionally, the information found on the following pages is intended for informational purposes only and should thus be considered, universal. As befitting its nature, the information presented is without assurance regarding its continued validity or interim quality. Trademarks that mentioned are done without written consent and can in no way be considered an endorsement from the trademark holder.

TABLE OF CONTENTS

DESCRIPTION 1

INTRODUCTION 2

INTRODUCTION: STOICISM AND THE VIRTUOUS LIFE 4

I.   WINTER WOES 7

II. SPRING IN BLOOM 45

III. BEAT THE HEAT WITH SUMMER VIRTUE 92

IV. FALL, A TIME OF CHANGE 117

BIBLIOGRAPHY 145

DESCRIPTION

This book is a collection of Stoic sayings organized to allow daily reference and inspiration. Including quotes from Marcus Aurelius, Seneca, Epictetus, and more, the Stoic advice covered in this volume runs the gambit from personal problems, to interpersonal relationships, to advice on work and productivity, to dealing with the hand of fate.

Face the world with a new light with the help of these immortal thinkers and learn both to conquer yourself and to come to terms with those things which you cannot control.

INTRODUCTION

Season One: Winter, Revival, and Coming to Terms.

a. Stoicism and Depression – Passages concerning sadness, coping, and revitalizing yourself.
b. Deprecation and Anger – How to handle changing moods, especially the strongest one.
c. Motivation – Stoic tips for staying active.
d. Fear, Regret and New Beginnings – Overcoming discrepancies between how you saw yourself last year and how you see yourself now.

Season Two: Spring, A Time for New Beginnings

a. Relinquishing the Past – Passages on forgiveness.
b. Body Image – Stoic advice for self-perception, focusing on control (or lack thereof) of one's appetites and health.
c. Life and Living Well – Stoic passages on birth and the promise of the future.
d. What to do With Work – Advice concerning heavy workloads and positive reinforcement.
e. Open to Possibilities – What the Stoics thought about views and anticipation of the future.

Summer: Problems in The Prime of Life

   a.   Relaxing: Not Just for Kids – The Stoics on leisure.
   b.   On Temperament and Temperature – Stoic advice on environmental and physical stress.
   c.   Living in Your Prime – How the Stoics viewed those in their prime and what to do with their vigor.

Fall: Change, Loss, and The End of Vitality

   a.   The Stoics and Loss – Stoics on dealing with life changes.
   b.   Being Prepared – Passages on reflecting on and preparing for shifts in fortune.
   c.   Dealing with Death – The Stoics and death, personal and impersonal.

Introduction: Stoicism and the Virtuous Life

Stoicism is an ancient philosophical school that has survived and thrived across ages, circumstances, and empires. Like many ancient schools, Stoicism has its origins in Athens. It first flourished alongside the noble Academy of Plato, the secretive Aristotelian Peripatetics, and the infamous Epicurean Garden. Stoicism's founder, Zeno of Citium, is famous for having taught freely in the "Stoa Poikile" or "Painted Porch" in English, from which the school derives its name.

The Stoics embody the love of wisdom. Their emphasis is practice, is living by example, both by teaching Stoic doctrine, particularly ethics, and by being exemplars of the doctrines they teach. Collectively, they define philosophy as a kind of activity, or *askêsis* in Greek, of knowledge concerning what is beneficial. Like the Epicureans, their approach to philosophy was therapeutic. They emphasized the development of good habits through knowledge of what is and is not to be valued. They aimed to strengthen the faculty of choice, *prohairesis* in Greek, and to thereby cultivate wisdom, to create Stoic sages.

The center and aim of the Stoic life is to be in accord with nature. Remember to keep separate their idea of nature from our modern idea of it. While it's true that instinct and inheritance plays a part in their concept, they also conceive of a thing's full development as belonging to its nature. If I ask "what is the nature of this seed?" you could say "to become a tree." This would be in accord with the Stoic idea of nature; it is not just the seed's nature to be an embryo housed in a coat with its nutrients, but also to grow into a tree in the right conditions. On the other hand, a person might be inclined to lie, cheat, and steal because of an evolutionary adaptation, and this may be the nature of an undeveloped person, but it is also within their nature to grow beyond that, to develop rationally and morally. Just as a seed that does not grow into a tree might be said to have failed with respect to its nature, so too can a person who does not develop morally be said to have failed.

The life cultivated in virtue will develop and mature morally. It is for this reason that the virtues must guide action in Stoic doctrine. The most important Stoic virtues are courage, temperance, prudence, and justice. For the Stoics, a life without these virtues is bestial, unbecoming of humanity. Further, these virtues are interdependent. One may be charitable in giving away a house to a friend, but that charity is not a virtue if the house given was stolen from someone else. Charity without justice, then, is in no way honorable. A fully realized human life is governed by virtue in its proper signification, as a well bound web worthy of the name wisdom.

I have in this book collected hundreds of quotes from various Stoic authors. The aim of this collection is to create a general resource or well of wisdom from which the reader can draw for their day-to-day trials. But alone these passages do not give a full picture of the Stoic project. One must fully account for, own up to, their nature as a creature predisposed toward virtue. This for the Stoics means consistent practice and self-reflection. It is not enough to, for example, observe what one can and cannot control when the impetus is disturbing or upsetting, but also when the impetus is self-affirming. It is not enough to say merely that an insult received from another person does not reflect one's own character, but also that a complement is so divorced.

This book is divide into four sections, each representing a season of the year. Besides finding this format aesthetically interesting, the organization of the themes is meant to carry a seasonal flare, with each set matching problems and attitudes often associated with the various seasons. Though the book itself starts in the winter theme, the reader is encouraged to skip around and find the theme that best suits her needs. By the end, it is my hope that she will come to understand Epictetus when he says,

"It has been ordained that there be summer and winter, abundance and dearth, virtue and vice, and all such opposites for the harmony of the whole, and (Zeus) has given each of us a body, property, and companions."

I. Winter Woes

a. **Stoicism and Sadness** – Passages concerning sadness, coping, and revitalizing yourself.

"Show me someone untroubled with disturbing thoughts about illness, danger, death, exile or loss of reputation. By all the gods, I want to see a Stoic!"

Epictetus, *Discourses*

"What really frightens and dismays us is not external events themselves, but the way in which we think about them. It is not things that disturb us, but our interpretation of their significance."

Epictetus, *Discourses*[1]

When you are disturbed by events and lose your serenity, quickly return to yourself and don't stay upset longer than the experience lasts; for you'll have more mastery over your inner harmony by continually returning to it.

---

1   Epictetus, *Discourses*, II, ibid

Marcus Aurelius, *Meditations*

"Fire tests gold, suffering tests brave men."

Seneca, *Letters From a Stoic*[2]

"Whatever happens, happens such as you are either formed by nature able to bear it, or not able to bear it. If such as you are by nature formed able to bear, bear it and fret not: But if such as you are not naturally able to bear, don't fret; for when it has consumed you, itself will perish. Remember, however, you are by nature formed able to bear whatever it is in the power of your own opinion to make supportable or tolerable, according as you conceive it advantageous, or your duty, to do so."

Marcus Aurelius, *Meditations*

"Be free from grief not through insensibility like the irrational animals, nor through want of thought like the foolish, but like a man of virtue by having reason as the consolation of grief."

---

2   Seneca, Letters *From a Stoic*, Letter I

Epictetus, Fragment[3]

Enough of this miserable, whining life. Stop monkeying around! Why are you troubled? What's new here? What's so confounding? The one responsible? Take a good look. Or just the matter itself? Then look at that. There's nothing else to look at. And as far as the gods go, by now you could try being more straightforward and kind. It's the same, whether you've examined these things for a hundred years, or only three.

Marcus Aurelius, *Meditations*

"Unhappy am I because this has happened to me.- Not so, but happy am I, though this has happened to me, because I continue free from pain, neither crushed by the present nor fearing the future."

Marcus Aurelius, *Meditations*

Failure to observe what is in the mind of another has seldom made a man unhappy; but those who do not observe the movements of their own minds must of necessity be unhappy.

---

[3] Epictetus, Fragment

Marcus Aurelius, *Meditations*[4]

"So let those people go on weeping and wailing whose self-indulgent minds have been weakened by long prosperity, let them collapse at the threat of the most trivial injuries; but let those who have spent all their years suffering disasters endure the worst afflictions with a brave and resolute staunchness. Everlasting misfortune does have one blessing, that it ends up by toughening those whom it constantly afflicts."

Seneca, *On Shortness of Life*[5]

"Faced with pain, you will discover the power of endurance. If you are insulted, you will discover patience. In time, you will grow to be confident that there is not a single impression that you will not have the moral means to tolerate."

Epictetus, *Enchiridion*[6]

Yes, keep on degrading yourself, soul. But soon your chance at dignity will be gone. Everyone gets one life. Yours is almost used up, and instead of treating yourself with respect, you have entrusted your own happiness to the souls of others.

---

[4]   Marcus Aurelius, *Meditations*, IV, III, ibid, II
[5]   Seneca, *On Shortness of Life*, II
[6]   Epictetus, *Enchiridion*, X

Marcus Aurelius, *Meditations*

"Wild animals run from the dangers they actually see, and once they have escaped them worry no more. We however are tormented alike by what is past and what is to come. A number of our blessings do us harm, for memory brings back the agony of fear while foresight brings it on prematurely. No one confines his unhappiness to the present."

Seneca, *Letters From a Stoic*[7]

Either all things proceed from one intelligent source and come together as in one body, and the part ought not to find fault with what is done for the benefit of the whole; or there are only atoms, and nothing else than mixture and dispersion. Why, then, are you disturbed?

Marcus Aurelius, *Meditations*[8]

"Your happiness depends on three things, all of which are within your power: your will, your ideas concerning the events in which you are involved, and the use you make of your ideas."

---

7   Seneca, *Letters From a Stoic*, Letter I
8   Marcus Aurelius, *Meditations*, V, IX

Epictetus, *Discourses*[9]

"Whenever you want to cheer yourself up, consider the good qualities of your companions, for example, the energy of one, the modesty of another, the generosity of yet another, and some other quality of another; for nothing cheers the heart as much as the images of excellence reflected in the character of our companions, all brought before us as fully as possible. Therefore, keep these images ready at hand."

Marcus Aurelius, *Meditations*

"Work, therefore to be able to say to every harsh appearance, 'You are but an appearance, and not absolutely the thing you appear to be.' And then examine it by those rules which you have, and first, and chiefly, by this: whether it concerns the things which are in our own control, or those which are not; and, if it concerns anything not in our control, be prepared to say that it is nothing to you."

Epictetus, *Enchiridion*

"If you suppose any of the things not in our own control to be either good or evil, when you are disappointed of what you wish, or incur what you would avoid, you must necessarily find fault with and blame

---

[9] Epictetus, *Discourses*, II

the authors. For every animal is naturally formed to fly and abhor things that appear hurtful, and the causes of them; and to pursue and admire those which appear beneficial, and the causes of them. It is impractical, then, that one who supposes himself to be hurt should be happy about the person who, he thinks, hurts him, just as it is impossible to be happy about the hurt itself."

Epictetus, *Enchiridion*[10]

"So long, in fact, as you remain in ignorance of what to aim at and what to avoid, what is essential and what is superfluous, what is upright or honorable conduct and what is not, it will not be traveling but drifting. All this hurrying from place to place won't bring you any relief, for you're traveling in the company of your own emotions, followed by your troubles all the way."

Seneca, *Letters From a Stoic*[11]

"The true man is revealed in difficult times. So when trouble comes, think of yourself as a wrestler whom God, like a trainer, has paired with a tough young buck. For what purpose? To turn you into Olympic-class material. But this is going to take some sweat to accomplish."

---

10   Epictetus, *Enchiridion*, I, XXXI
11   Seneca, *Letters From a Stoic*, Letter I

Epictetus, *Discourses*

"Remember from now on whenever something tends to make you unhappy, draw on this principle: 'This is no misfortune; but bearing with it bravely is a blessing.'"

Epictetus, *Discourses*[12]

b. **Deprecation and Anger** – How to handle changing moods, especially the strongest one.

"Remember, it is not enough to be hit or insulted to be harmed, you must believe that you are being harmed. If someone succeeds in provoking you, realize that your mind is complicit in the provocation. Which is why it is essential that we not respond impulsively to impressions; take a moment before reacting, and you will find it easier to maintain control."

Epictetus, *Discourses*[13]

"Anger cannot be dishonest."

---

12 Epictetus, *Discourses*, II, IV
13 Epictetus, *Discourses*, I

Marcus Aurelius, *Meditations*

"A physician is not angry at the intemperance of a mad patient; nor does he take it ill to be railed at by a man in a fever. Just so should a wise man treat all mankind, as a physician does his patient; and looking upon them only as sick and extravagant."

Seneca, *Letters From a Stoic*[14]

It is a proper work of a man to be benevolent to his own kind, to despise the movements of the senses, to form a just judgment of plausible appearances, and to take a survey of the nature of the universe and of the things that happen in it.

Marcus Aurelius, *Meditations*

"Men decide far more problems by hate, love, lust, rage, sorrow, joy, hope, fear, illusion, or some other inward emotion than by reality, authority, any legal standard, judicial precedent, or statute."

---

[14] Seneca, *Letters From a Stoic*, Letter II

Cicero, *De Oratore*[15]

Consider what men are when they are eating, sleeping, coupling, evacuating, and so forth. Then what kind of men they are when they are imperious and arrogant, or angry and scolding from their elevated place.

Marcus Aurelius, *Meditations*

"An innocent man, if accused, can be acquitted; a guilty man, unless accused, cannot be condemned. It is, however, more advantageous to absolve an innocent than not to prosecute a guilty man."

Cicero, *On Anger*[16]

"As for others whose lives are not so ordered, he reminds himself constantly of the characters they exhibit daily and nightly at home and abroad, and of the sort of society they frequent; and the approval of such men, who do not even stand well in their own eyes, has no value for him."

---

15   Cicero, *De Oratore*
16   Cicero, *On Anger*, II

Marcus Aurelius, *Meditations*[17]

"As fire when thrown into water is cooled down and put out, so also a false accusation when brought against a man of the purest and holiest character, boils over and is at once dissipated, and vanishes and threats of heaven and sea, himself standing unmoved."

Cicero, *On Anger*[18]

"Nothing is burdensome if taken lightly, and nothing need arouse one's irritation so long as one doesn't make it bigger than it is by getting irritated."

Seneca, *Letters From a Stoic*

This, then, is consistent with the character of a reflecting man, to be neither careless nor impatient nor contemptuous with respect to death, but to wait for it as one of the operations of nature.

Marcus Aurelius, *Meditations*

'My brother shouldn't have treated me in this way.' Indeed he shouldn't, but it's for him to see to that. For my part, however he treats me, I should conduct myself towards him as I ought. For that is my

---

17 Marcus Aurelius, *Meditations*, II, VI, ibid, III
18 Cicero, *On Anger*, I

business, and the rest is not my concern. In this no one can hinder me, while everything else is subject to hindrance."

Epictetus, *Discourses*[19]

Consider that you also do many things wrong, and that you are a man like others; and even if you do abstain from certain faults, still you have the disposition to commit them, though either through cowardice, or concern about reputation, or some such mean motive, you abstain from such faults.

Marcus Aurelius, *Meditations*[20]

"Treat unenlightened souls with sympathy and indulgence, remembering that they are ignorant or mistaken about what's most important. Never be harsh, remember Plato's dictum: 'Every soul is deprived of the truth against its will.'"

Epictetus, *Discourses*

"If you really want to escape the things that harass you, what you're needing is not to be in a different place but to be a different person."

---

19   Epictetus, *Discourses*, III, II
20   Marcus Aurelius, *Meditations*, IX, IV

Seneca, *Letters From a Stoic*[21]

Whatever any one does or says, I must be good, just as if the emerald (or the gold or the purple) were always saying "Whatever any one does or says, I must be emerald and keep my color."

Marcus Aurelius, *Meditations*

If you don't want to be cantankerous, don't feed your temper, or multiply incidents of anger. Suppress the first impulse to be angry, then begin to count the days on which you don't get mad.

Marcus Aurelius, *Meditations*

"He did not say, 'Define me envy', and then, when the man defined it, 'You define it ill, for the terms of the definition do not correspond to the subject defined.' Such phrases are technical and therefore tiresome to the lay mind, and hard to follow, yet you and I cannot get away from them. We are quite unable to rouse the ordinary man's attention in a way which will enable him to follow his own impressions and so arrive at admitting or rejecting this or that. And therefore those of us who are at all cautious naturally give the subject up, when we become aware of this incapacity; while the mass of men, who venture at random into this sort of enterprise, muddle others and get muddled themselves, and end

---

21 Seneca, *Letters From a Stoic*, Letter II, ibid

by abusing their opponents and getting abused in return, and so leave the field. But the first quality of all in Socrates, and the most characteristic, was that he never lost his temper in argument, never uttered anything abusive, never anything insolent, but bore with abuse from others and quieted strife."

Epictetus, *Enchiridion*[22]

Who wants to live with delusion and prejudice, being unjust, undisciplined, mean and ungrateful? 'No one.' No bad person, then, lives the way he wants, and no bad man is free.

Marcus Aurelius, *Meditations*

"For what does the man who accepts insult do that is wrong? It is the doer of wrong who puts themselves to shame-the sensible man wouldn't go to the law, since he wouldn't even consider that he had been insulted! Besides, to be annoyed or angered about such things would be petty-instead easily and silently bear what has happened, since this is appropriate for those whose purpose is to be noble-minded."

---

22 Epictetus, *Enchiridion*

Musonius Rufus, *How To Live*[23]

"If any man despises me, that is his problem. My only concern is not doing or saying anything deserving of contempt."

Marcus Aurelius, *Meditations*[24]

"....nothing cruel is in fact beneficial; for cruelty is extremely hostile to the nature of man, which we ought to follow."

Cicero, *On Anger*[25]

With what are you discontented? With the badness of men? Recall to your mind this conclusion, that rational animals exist for one another, and that to endure is a part of justice, and that men do wrong involuntarily.

Marcus Aurelius, *Meditations*

"When you wake up in the morning, tell yourself: the people I deal with today will be meddling, ungrateful, arrogant, dishonest, jealous and surly. They are like this because they can't tell good from evil. But I have seen the beauty of good, and the ugliness of evil, and have recognized that the wrongdoer has a nature related to my own - not of

---

23 Musonius Rufus, *How To Live*
24 Marcus Aurelius, *Meditations*, VII, II, III
25 Cicero, *On Anger*

the same blood and birth, but the same mind, and possessing a share of the divine. And so none of them can hurt me. No one can implicate me in ugliness. Nor can I feel angry at my relative, or hate him. We were born to work together like feet, hands and eyes, like the two rows of teeth, upper and lower. To obstruct each other is unnatural. To feel anger at someone, to turn your back on him: these are unnatural."

Marcus Aurelius, *Meditations*

"To accuse others for one's own misfortune is a sign of want of education. To accuse oneself shows that one's education has begun. To accuse neither oneself nor others shows that one's education is complete."

Epictetus, *Enchiridion*[26]

"How much more grievous are the consequences of anger than the causes of it."

Marcus Aurelius, *Meditations*

"Men seek retreats for themselves, houses in the country, sea-shores, and mountains; and thou too art wont to desire such things very much.

---

26   Epictetus, *Enchiridion*, V

But this is altogether a mark of the most common sort of men, for it is in thy power whenever thou shalt choose to retire into thyself. For nowhere either with more quiet or more freedom from trouble does a man retire than into his own soul, particularly when he has within him such thoughts that by looking into them he is immediately in perfect tranquility; and I affirm that tranquility is nothing else than the good ordering of the mind."

Marcus Aurelius, *Meditations*[27]

"What, for instance, does it mean to be insulted? Stand by a rock and insult it, and what have you accomplished? If someone responds to insult like a rock, what has the abuser gained with his invective? If, however, he has his victim's weakness to exploit, then his efforts are worth his while."

Epictetus, *Of Human Freedom*[28]

When a guide meets up with someone who is lost, ordinarily his reaction is to direct him on the right path, not mock or malign him, then turn on his heel and walk away. As for you, lead someone to the truth and you will find that he can follow. But as long as you don't

---

27 Marcus Aurelius, *Meditations,* IV, II, I, IV
28 Epictetus, *Of Human Freedom*

point it out to him, don't make fun of him; be aware of what you need to work on instead.

Marcus Aurelius, *Meditations*

"Some things are in our control and others not. Things in our control are opinion, pursuit, desire, aversion, and, in a word, whatever are our own actions. Things not in our control are body, property, reputation, command, and, in one word, whatever are not our actions. The things in our control are by nature free, unrestrained, unhindered; but those not in our control are weak, slavish, restrained, belonging to others. Remember, then, that if you suppose that things which are slavish by nature are also free, and that what belongs to others is your own, then you will be hindered. You will lament, you will be disturbed, and you will find fault both with gods and men. But if you suppose that only to be your own which is your own, and what belongs to others such as it really is, then no one will ever compel you or restrain you. Further, you will find fault with no one or accuse no one. You will do nothing against your will. No one will hurt you, you will have no enemies, and you not be harmed."

Epictetus, *Enchiridion*[29]

"The best revenge is to be unlike him who performed the injury."

Marcus Aurelius, *Meditations*

"You always own the option of having no opinion. There is never any need to get worked up or to trouble your soul about things you can't control. These things are not asking to be judged by you. Leave them alone."

Marcus Aurelius, *Meditations*

"Regain your senses, call yourself back, and once again wake up. Now that you realize that only dreams were troubling you, view this 'reality' as you view your dreams."

Marcus Aurelius, *Meditations*[30]

"When people injure you, ask yourself what good or harm they thought would come of it. If you understand that, you'll feel sympathy rather than outrage or anger. Your sense of good and evil may be the same as theirs, or near it, in which case you have to excuse them. Or

---

29 Epictetus, *Enchiridion*, I
30 Marcus Aurelius, *Meditations*, II, I, ibid, V, IV

your sense of good and evil may differ from theirs. In which case they're misguided and deserve your compassion. Is that so hard?"

Marcus Aurelius, *Meditations*

c. **Motivation** – Stoic tips for staying active.

You have leisure or ability to check arrogance: you have leisure to be superior to pleasure and pain: you have leisure to be superior to love of fame, and not to be vexed at stupid and ungrateful people, nay even to care for them.

Marcus Aurelius, *Meditations*

"Your greatest difficulty is with yourself; for you are your own stumbling-block. You do not know what you want. You are better at approving the right course than at following it out. You see where the true happiness lies, but you have not the courage to attain it."

Seneca, *Letters From a Stoic*[31]

Nothing important comes into being overnight; even grapes or figs need time to ripen. If you say that you want a fig now, I will tell you to be patient. First, you must allow the tree to flower, then put forth fruit;

---

[31]   Seneca, *Letters From a Stoic*, Letter 21

then you have to wait until the fruit is ripe. So if the fruit of a fig tree is not brought to maturity instantly or in an hour, how do you expect the human mind to come to fruition, so quickly and easily?

Epictetus, *Discourses*

"It is not reasonings that are wanted now,' he says, 'for there are books stuffed full of stoical reasonings. What is wanted, then? The man who shall apply them; whose actions may bear testimony to his doctrines. Assume this character for me, that we may no longer make use in the schools of the examples of the ancients, but may have some examples of our own."

Epictetus, *Discourses*[32]

"In your actions, don't procrastinate. In your conversations, don't confuse In your thoughts, don't wander. In your soul, don't be passive or aggressive. In your life, don't be all about business."

---

[32]   Epictetus, *Discourses*, II, I

Marcus Aurelius, *Meditations*[33]

"When you're called upon to speak, then speak, but never about banalities like gladiators, horses, sports, food and drink – commonplace stuff. Above all don't gossip about people, praising, blaming or comparing them."

Marcus Aurelius, *Meditations*

"Continue to act thus, my dear Lucilius – set yourself free for your own sake; gather and save your time, which till lately has been forced from you, or filched away, or has merely slipped from your hands. Make yourself believe the truth of my words, – that certain moments are torn from us, that some are gently removed, and that others glide beyond our reach. The most disgraceful kind of loss, however, is that due to carelessness. Furthermore, if you will pay close heed to the problem, you will find that the largest portion of our life passes while we are doing ill, a goodly share while we are doing nothing, and the whole while we are doing that which is not to the purpose. What man can you show me who places any value on his time, who reckons the worth of each day, who understands that he is dying daily? For we are mistaken when we look forward to death; the major portion of death has already passed. Whatever years be behind us are in death's hands."

---

33   Marcus Aurelius, *Meditations*, VIII, IV, II

Seneca, *Letters From a Stoic*[34]

"Set yourself in motion, if it is in your power, and do not look about you to see if anyone will observe it; nor yet expect Plato's Republic: but be content if the smallest thing goes on well, and consider such an event to be no small matter.'

Marcus Aurelius, *Meditations*[35]

"What then, is it not possible to be free from faults? It is not possible; but this is possible: to direct your efforts incessantly to being faultless. For we must be content if by never remitting this attention we shall escape at least a few errors. When you have said "Tomorrow I will begin to attend," you must be told that you are saying this: "Today I will be shameless, disregardful of time and place, mean;it will be in the power of others to give me pain, today I will be passionate and envious.

See how many evil things you are permitting yourself to do. If it is good to use attention tomorrow, how much better is it to do so today? If tomorrow it is in your interest to attend, much more is it today, that you may be able to do so tomorrow also, and may not defer it again to the third day."

---

34   Seneca, *Letters From a Stoic*, Letter 10, Letter 13
35   Marcus Aurelius, *Meditations*, XII, III

Epictetus, *Discourses*

"Nothing important comes into being overnight; even grapes and figs need time to ripen. If you say that you want a fig now, I will tell you to be patient. First, you must allow the tree to flower, then put forth fruit; then you have to wait until the fruit is ripe. So if the fruit of a fig tree is not brought to maturity instantly or in an hour, how do you expect the human mind to come to fruition, so quickly and easily?"

Epictetus, *Discourses* [36]

"It is in your power to live here. But if men do not permit you, then get away out of life, as if you were suffering no harm. The house is smoky, and I quit it. Why do you think that this is any trouble? But so long as nothing of the kind drives me out, I remain, am free, and no man shall hinder me from doing what I choose; and I choose to do what is according to the nature of the rational and social animal."

Marcus Aurelius, *Meditations*

"I never spend a day in idleness; I appropriate even a part of the night for study. I do not allow time for sleep but yield to it when I must, and when my eyes are wearied with waking and ready to fall shut, I keep them at their task."

---

[36] Epictetus, *Discourses*, IV

Seneca, *Letters From a Stoic*

"And so there is no reason for you to think that any man has lived long because he has grey hairs or wrinkles, he has not lived long – he has existed long. For what if you should think that man had had a long voyage who had been caught by a fierce storm as soon as he left harbour, and, swept hither and thither by a succession of winds that raged from different quarters, had been driven in a circle around the same course? Not much voyaging did he have, but much tossing about."

Seneca, *On Shortness of Life*[37]

"At dawn, when you have trouble getting out of bed, tell yourself, "I have to go to work - as a human being. What do I have to complain of, if I'm going to do what I was born for - the things I was brought into the world to do? Or is this what I was created for? To huddle under the blankets and stay warm?"

Marcus Aurelius, *Meditations*[38]

"It is not that we have a short time to live, but that we waste a lot of it. Life is long enough, and a sufficiently generous amount has been given to us for the highest achievements if it were all well invested. But when

---

37   Seneca, *On Shortness of Life*
38   Marcus Aurelius, *Meditations*, V, I

it is wasted in heedless luxury and spent on no good activity, we are forced at last by death's final constraint to realize that it has passed away before we knew it was passing. So it is: we are not given a short life but we make it short, and we are not ill-supplied but wasteful of it... Life is long if you know how to use it."

Seneca, *Letters From a Stoic*[39]

"Most of what passes for legitimate entertainment is inferior or foolish and only caters to or exploits people's weaknesses. Avoid being one of the mob who indulges in such pastimes. Your life is too short and you have important things to do. Be discriminating about what images and ideas you permit into your mind. If you yourself don't choose what thoughts and images you expose yourself to, someone else will, and their motives may not be the highest. It is the easiest thing in the world to slide imperceptibly into vulgarity. But there's no need for that to happen if you determine not to waste your time and attention on mindless pap."

Epictetus, *Discourses*[40]

"Concentrate every minute like a Roman— like a man— on doing what's in front of you with precise and genuine seriousness, tenderly,

---

39 Seneca, *Letters From a Stoic*, Letter 12
40 Epictetus, *Discourses*, IV

willingly, with justice. And on freeing yourself from all other distractions. Yes, you can— if you do everything as if it were the last thing you were doing in your life, and stop being aimless, stop letting your emotions override what your mind tells you, stop being hypocritical, self-centered, irritable. You see how few things you have to do to live a satisfying and reverent life? If you can manage this, that's all even the gods can ask of you."

Marcus Aurelius, *Meditations*

"The chief task in life is simply this: to identify and separate matters so that I can say clearly to myself which are externals not under my control, and which have to do with the choices I actually control. Where then do I look for good and evil? Not to uncontrollable externals, but within myself to the choices that are my own..."

Epictetus, *Discourses*[41]

"If anyone says that the best life of all is to sail the sea, and then adds that I must not sail upon a sea where shipwrecks are a common occurrence and there are often sudden storms that sweep the helmsman in an adverse direction, I conclude that this man, although he lauds navigation, really forbids me to launch my ship."

---

41   Epictetus, *Discourses*, II

Seneca, *On The Shortness of Life*

"Putting things off is the biggest waste of life: it snatches away each day as it comes, and denies us the present by promising the future. The greatest obstacle to living is expectancy, which hangs upon tomorrow, and loses today. You are arranging what lies in Fortune's control, and abandoning what lies in yours. What are you looking at? To what goal are you straining? The whole future lies in uncertainty: live immediately."

Seneca, *On The Shortness of Life*[42]

"Look well into thyself; there is a source of strength which will always spring up if thou wilt always look."

Marcus Aurelius, *Meditations*[43]

"How long are you going to wait before you demand the best for yourself and in no instance bypass the discrimination of reason? You have been given the principles that you ought to endorse, and you have endorsed them. What kind of teacher, then, are you still waiting for in order to refer your self-improvement to him? You are no longer a boy, but a full-grown man. If you are careless and lazy now and keep

---

42   Seneca, *On The Shortness of Life*
43   Marcus Aurelius, *Meditations,* I, ibid

putting things off and always deferring the day after which you will attend to yourself, you will not notice that you are making no progress, but you will live and die as someone quite ordinary."

Epictetus, *Enchiridion*[44]

'Stop wandering about! You aren't likely to read your own notebooks, or ancient histories, or the anthologies you've collected to enjoy in your old age. Get busy with life's purpose, toss aside empty hopes, get active in your own rescue-if you care for yourself at all-and do it while you can."

Marcus Aurelius, *Meditations*

"As the sun does not wait for prayers and incantations to be induced to rise, but immediately shines and is saluted by all, so do you also not wait for clapping of hands and shouts of praise to be induced to do good, but be a doer of good voluntarily and you will be beloved as much as the sun."

---

44   Epictetus, *Enchiridion*, 22

Epictetus, *Discourses*

"A noble man compares and estimates himself by an idea which is higher than himself; and a mean man, by one lower than himself. The one produces aspiration; the other ambition, which is the way in which a vulgar man aspires."

Marcus Aurelius, *Meditations*[45]

"You will do the greatest services to the state, if you shall raise not the roofs of the houses, but the souls of the citizens: for it is better that great souls should dwell in small houses than for mean slaves to lurk in great houses."

Epictetus, *Enchiridion*[46]

"What would have become of Hercules do you think if there had been no lion, hydra, stag or boar - and no savage criminals to rid the world of? What would he have done in the absence of such challenges?

Obviously he would have just rolled over in bed and gone back to sleep. So by snoring his life away in luxury and comfort he never would have developed into the mighty Hercules.

---

45   Marcus Aurelius, *Meditations*, V
46   Epictetus, *Enchiridion*

And even if he had, what good would it have done him? What would have been the use of those arms, that physique, and that noble soul, without crises or conditions to stir into him action?"

Epictetus, *Discourses*[47]

"Can anything be more idiotic than certain people who boast of their foresight? They keep themselves officiously preoccupied in order to improve their lives; they spend their lives in organizing their lives. They direct their purposes with an eye to a distant future. But putting things off is the biggest waste of life: it snatches away each day as it comes, and denies us the present by promising the future. The greatest obstacle to living is expectancy, which hangs upon tomorrow and loses today. You are arranging what lies in Fortune's control, and abandoning what lies in yours. What are you looking at? To what goal are you straining?"

Seneca, *Letters From a Stoic*[48]

"If you have an earnest desire towards philosophy, prepare yourself from the very first to have the multitude laugh and sneer, and say, "He is returned to us a philosopher all at once; "and "Whence this supercilious look?" Now, for your part, do not have a supercilious look

---

47   Epictetus, *Discourses*, II, ibid
48   Seneca, *Letters From a Stoic*, Letter 20

indeed; but keep steadily to those things which appear best to you, as one appointed by God to this particular station. For remember that, if you are persistent, those very persons who at first ridiculed will afterwards admire you. But if you are conquered by them, you will incur a double ridicule."

Epictetus, *Enchiridion*[49]

d. **Fear, Regret and New Beginnings** – Overcoming discrepancies between how you saw yourself last year and how you see yourself now.

Don't waste the rest of your time here worrying about other people – unless it affects the common good. It will keep you from doing anything useful. You'll be too preoccupied with what so-and-so is doing, and why, and what they're saying, and what they're thinking, and what they're up to, and all the other things that throw you off and keep you from focusing on your own mind.

Marcus Aurelius, *Meditations*

"Limiting one's desires actually helps to cure one of fear. 'Cease to hope … and you will cease to fear.' … Widely different [as fear and

---

49   Epictetus, *Enchiridion*

hope] are, the two of them march in unison like a prisoner and the escort he is handcuffed to. Fear keeps pace with hope … both belong to a mind in suspense, to a mind in a state of anxiety through looking into the future. Both are mainly due to projecting our thoughts far ahead of us instead of adapting ourselves to the present."

Seneca, *Letters From a Stoic*[50]

"Today I escaped anxiety. Or no, I discarded it, because it was within me, in my own perceptions — not outside."

Marcus Aurelius, *Meditations*[51]

"Fear is the cause -not exile. To many people, even to most, despite living safely in their home city, fear of what seem to them the dire consequences of free speech is present. The courageous, in exile or at home, is fearless in the face of all such threats; for that reason they've the courage to say what they think equally at home or in exile."

---

50 Seneca, *Letters From a Stoic*, Letter 3
51 Marcus Aurelius, *Meditations*, V, II, V

Musonius Rufus, *How To Live*[52]

"When force of circumstance upsets your equanimity, lose no time in recovering your self-control, and do not remain out of tune longer than you can help. Habitual recurrence to the harmony will increase your mastery of it."

Marcus Aurelius, Meditations

"But is life really worth so much? Let us examine this; it's a different inquiry. We will offer no solace for so desolate a prison house; we will encourage no one to endure the over lordship of butchers. We shall rather show that in every kind of slavery, the road of freedom lies open. I will say to the man to whom it befell to have a king shoot arrows at his dear ones, and to him whose master makes fathers banquet on their sons' guts: 'What are you groaning for, fool?... Everywhere you look you find an end to your sufferings. You see that steep drop-off? It leads down to freedom. You see that ocean, that river, that well? Freedom lies at its bottom. You see that short, shriveled, bare tree? Freedom hangs from it.... You ask, what is the path to freedom? Any vein in your body.'"

---

52   Musonius Rufus, *How To Live*

Seneca, *Letters From a Stoic*

"So what oppresses and scares us? It is our own thoughts, obviously, What overwhelms people when they are about to leaves friends, family, old haunts and their accustomed way of life? Thoughts."

Epictetus, *Discourses*[53]

"Live a good life. If there are gods and they are just, then they will not care how devout you have been, but will welcome you based on the virtues you have lived by. If there are gods, but unjust, then you should not want to worship them. If there are no gods, then you will be gone, but will have lived a noble life that will live on in the memories of your loved ones."

Marcus Aurelius, *Meditations*[54]

"The trip doesn't exist that can set you beyond the reach of cravings, fits of temper, or fears … so long as you carry the sources of your troubles about with you, those troubles will continue to harass and plague you wherever you wander on land or on sea. Does it surprise you that running away doesn't do you any good? The things you're running away from are with you all the time."

---

53   Epictetus, *Discourses*, II
54   Marcus Aurelius, *Meditations*, III, I

Seneca, *Letters From a Stoic*[55]

"As for us, we behave like a herd of deer. When they flee from the huntsman's feathers in affright, which way do they turn? What haven of safety do they make for? Why, they rush upon the nets! And thus they perish by confounding what they should fear with that wherein no danger lies. . . . Not death or pain is to be feared, but the fear of death or pain. Well said the poet therefore:—

Death has no terror; only a Death of shame!"

Epictetus, *Fragments*[56]

"Do not waste what remains of your life in speculating about your neighbors, unless with a view to some mutual benefit. To wonder what so-and-so is doing and why, or what he is saying, or thinking, or scheming—in a word, anything that distracts you from fidelity to the ruler within you—means a loss of opportunity for some other task."

---

55   Seneca, *Letters From a Stoic*, Letter 24
56   Epictetus, Fragment

Marcus Aurelius, *Meditations*

"The gods do not exists, and even if they exist they do not trouble themselves about people, and we have nothing in common with them. The piety and devotion to the gods that the majority of people invoke is a lie devised by swindlers and con men and, if you can believe it, by legislators, to keep criminals in line by putting the fear of God into them."

Epictetus, *Discourses*

"Do not disturb yourself by picturing your life as a whole; do not assemble in your mind the many and varied troubles which have come to you in the past and will come again in the future, but ask yourself with regard to every present difficulty: 'What is there in this that is unbearable and beyond endurance?' You would be ashamed to confess it! And then remind yourself that it is not the future or what has passed that afflicts you, but always the present, and the power of this is much diminished if you take it in isolation and call your mind to task if it thinks that it cannot stand up to it when taken on its own."

Marcus Aurelius, *Meditations*[57]

"With every accident, ask yourself what abilities you have for making a proper use of it. If you see an attractive person, you will find that self-restraint is the ability you have against your desire. If you are in pain, you will find fortitude. If you hear unpleasant language, you will find patience. And thus habituated, the appearances of things will not hurry you away along with them."

Epictetus, *Enchiridion*[58]

"Two elements must therefore be rooted out once for all, – the fear of future suffering, and the recollection of past suffering; since the latter no longer concerns me, and the former concerns me not yet."

Seneca, *Letters From a Stoic*[59]

---

[57] Marcus Aurelius, *Meditations*, II
[58] Epictetus, *Enchiridion*
[59] Seneca, *Letters From a Stoic*, Letter 78

II. Spring in Bloom

a. **Relinquishing the Past** – Passages on forgiveness.

"Philosophy does not promise to secure anything external for man, otherwise it would be admitting something that lies beyond its proper subject-matter. For as the material of the carpenter is wood, and that of statuary bronze, so the subject-matter of the art of living is each person's own life."

Epictetus, *Discourses*[60]

"You have the power to strip away many superfluous troubles located wholly in your judgment, and to possess a large room for yourself embracing in thought the whole cosmos, to consider everlasting time, to think of the rapid change in the parts of each thing, of how short it is from birth until dissolution, and how the void before birth and that after dissolution are equally infinite."

Marcus Aurelius, *Meditations*

"To see a man fearless in dangers, untainted with lusts, happy in adversity, composed in a tumult, and laughing at all those things which are generally either coveted or feared, all men must acknowledge that

---

[60] Epictetus, *Discourses*, I, III

this can be from nothing else but a beam of divinity that influences a mortal body."

Seneca, *Letters From a Stoic*

"To admonish is better than to reproach for admonition is mild and friendly, but reproach is harsh and insulting; and admonition corrects those who are doing wrong, but reproach only convicts them."

Epictetus, Enchiridion[61]

"Whenever you are about to find fault with someone, ask yourself the following question: What fault of mine most nearly resembles the one I am about to criticize?"

Marcus Aurelius, *Meditations*[62]

"No man has ever been so far advanced by Fortune that she did not threaten him as greatly as she had previously indulged him. Do not trust her seeming calm; in a moment the sea is moved to its depths. The very day the ships have made a brave show in the games, they are engulfed."

---

61 Epictetus, *Enchiridion*
62 Marcus Aurelius, *Meditations*, V, VII

Seneca, *Letters From a Stoic*[63]

"Does anyone bathe hastily? Do not say that they do it ill, but hastily. Does anyone drink much wine? Do not say that they do ill, but that they drink a great deal. For unless you perfectly understand their motives, how should you know if they act ill? Thus you will not risk yielding to any appearances except those you fully comprehend."

Epictetus, *Enchiridion*[64]

It is just charming how people boast about qualities beyond their control. For instance, 'I am better than you because I have many estates, while you are practically starving'; or, 'I'm a consul,' 'I'm a governor,' or 'I have fine curly hair.'

Marcus Aurelius, *Meditations*

"To help us to cheerfully endure those hardships which we may expect to suffer because of virtue and goodness, it is useful to recall what hardships people will endure for immoral reasons. Consider what lustful lovers undergo for the sake of evil desires-and how much exertion others expend for the sake of profit-how much suffering pursuing fame - bear in mind that they all submit to all kinds of toil

---

63   Seneca, *Letters From a Stoic*, Letter 57
64   Epictetus, *Enchiridion*, X

and hardship voluntarily. It's monstrous that they endure such things for no honourable reward, yet for the sake of the good (not only the avoidance of evil that wrecks our lives-also the gain of virtue) we're not ready to bear the slightest hardship."

Musonius Rufus, *On How To Live*[65]

"You should be especially careful when associating with one of your former friends or acquaintances not to sink to their level; otherwise you will lose yourself. If you are troubled by the idea that 'He'll think I'm boring and won't treat me the way he used to,' remember that everything comes at a price. It isn't possible to change your behavior and still be the same person you were before."

Epictetus, *Discourses*

b. **Self-Image** – Stoic advice for self-perception, focusing on control (or lack thereof) of one's appetites and health.

"You know yourself what you are worth in your own eyes; and at what price you will sell yourself. For men sell themselves at various prices. This is why, when Florus was deliberating whether he should appear at Nero's shows, taking part in the performance himself,

---

65   Musonius Rufus, *On How To Live*

Agrippinus replied, 'Appear by all means.' And when Florus inquired, 'But why do not you appear?' he answered, 'Because I do not even consider the question.' For the man who has once stooped to consider such questions, and to reckon up the value of external things, is not far from forgetting what manner of man he is."

Epictetus, *Discourses*[66]

"Because your own strength is unequal to the task, do not assume that it is beyond the powers of man; but if anything is within the powers and province of man, believe that it is within your own compass also."

Marcus Aurelius, *Meditations*[67]

"Why be concerned about others, come to that, when you've outdone your own self? Set yourself a limit which you couldn't even exceed if you wanted to, and say good-bye at last to those deceptive prizes more precious to those who hope for them than to those who have won them. If there were anything substantial in them they would sooner or later bring a sense of fullness; as it is they simply aggravate the thirst of those who swallow them."

---

66   Epictetus, Discourses, IV, II
67   Marcus Aurelius, *Meditations*, III

Seneca, *Letters From a Stoic*[68]

"You ought to realize, you take up very little space in the world as a whole – your body, that is; in reason, however, you yield to no one, not even to the gods, because reason is not measured in size but sense. So why not care for that side of you, where you and the gods are equals?"

Marcus Aurelius, *Meditations*

"I believe that no characteristic is so distinctively human as the sense of indebtedness we feel, not necessarily for a favor received, but even for the slightest evidence of kindness; and there is nothing so boorish, savage, inhuman as to appear to be overwhelmed by a favor, let alone unworthy of it."

Cicero, *On Old Age*

"When you have done a good act and another has received it, why do you look for a third thing besides these, as fools do, either to have the reputation of having done a good act or to obtain a return? If money is your only standard, then consider that, by your lights, someone who loses their nose does not suffer any harm."

---

[68] Seneca, *Letters From a Stoic*, Letter 33

Marcus Aurelius, *Meditations*

We are motivated by a keen desire for praise, and the better a man is the more he is inspired by glory. The very philosophers themselves, even in those books which they write in contempt of glory, inscribe their names.

Cicero, *On Old Age*[69]

"Never depend on the admiration of others. There is no strength in it. Personal merit cannot be derived from an external source. It is not to be found in your personal associations, nor can it be found in the regard of other people. It is a fact of life that other people, even people who love you, will not necessarily agree with your ideas, understand you, or share your enthusiasms. Grow up! Who cares what other people think about you!"

Epictetus, *Enchiridion*

"How strangely men act. They will not praise those who are living at the same time and living with themselves; but to be themselves praised by posterity, by those whom they have never seen or ever will see, this they set much value on."

---

69   Cicero, *On Old Age*

Marcus Aurelius, *Meditations*

"Do as Socrates did, never replying to the question of where he was from with, 'I am Athenian,' or 'I am from Corinth,' but always, 'I am a citizen of the world.'"

Marcus Aurelius, *Meditations*[70]

"I imagine many people could have achieved wisdom if they had not imagined they had already achieved it, if they had not dissembled about some of their own characteristics and turned a blind eye to others."

Seneca, *Letters From a Stoic*[71]

"So you know how things stand. Now forget what they think of you. Be satisfied if you can live the rest of your life, however short, as your nature demands. Focus on that, and don't let anything distract you. You've wandered all over and finally realized that you never found what you were after: how to live. Not in syllogisms, not in money, or fame, or self-indulgence. Nowhere."

---

70   Marcus Aurelius, *Meditations*, II, IV, ibid
71   Seneca, *Letters From a Stoic*, Letter 28

Marcus Aurelius, *Meditations*

"A person's worth is measured by the worth of what he values."

Marcus Aurelius, *Meditations*

"When any person harms you, or speaks badly of you, remember that he acts or speaks from a supposition of its being his duty. Now, it is not possible that he should follow what appears right to you, but what appears so to himself. Therefore, if he judges from a wrong appearance, he is the person hurt, since he too is the person deceived. For if anyone should suppose a true proposition to be false, the proposition is not hurt, but he who is deceived about it. Setting out, then, from these principles, you will meekly bear a person who reviles you, for you will say upon every occasion, "It seemed so to him.""

Epictetus, *Enchiridion*[72]

"The happiness of those who want to be popular depends on others; the happiness of those who seek pleasure fluctuates with moods outside their control; but the happiness of the wise grows out of their own free acts."

---

[72] Epictetus, *Enchiridion*

Marcus Aurelius, *Meditations*[73]

"-Who are those people by whom you wish to be admired? Are they not these whom you are in the habit of saying that they are mad? What then? Do you wish to be admired by the mad?"

Epictetus, *Discourses*

"Never value anything as profitable that compels you to break your promise, to lose your self-respect, to hate any man, to suspect, to curse, to act the hypocrite, to desire anything that needs walls and curtains."

Marcus Aurelius, *Meditations*

"Crows pick out the eyes of the dead, when the dead have no longer need of them; but flatterers mar the soul of the living, and her eyes they blind."

Epictetus, *Discourses*[74]

"In your conversation, don't dwell at excessive length on your own deeds or adventures. Just because you enjoy recounting your exploits

---

73   Marcus Aurelius, *Meditations,* IV, ibid, II
74   Epictetus, *Discourses,* II, ibid

doesn't mean that others derive the same pleasure from hearing about them."

Marcus Aurelius, *Meditations*[75]

"And this, too, affords no small occasion for anxieties - if you are bent on assuming a pose and never reveal yourself to anyone frankly, in the fashion of many who live a false life that is all made up for show; for it is torturous to be constantly watching oneself and be fearful of being caught out of our usual role. And we are never free from concern if we think that every time anyone looks at us he is always taking-our measure; for many things happen that strip off our pretense against our will, and, though all this attention to self is successful, yet the life of those who live under a mask cannot be happy and without anxiety. But how much pleasure there is in simplicity that is pure, in itself unadorned, and veils no part of its character! Yet even such a life as this does run some risk of scorn, if everything lies open to everybody; for there are those who disdain whatever has become too familiar. But neither does virtue run any risk of being despised when she is brought close to the eyes, and it is better to be scorned by reason of simplicity than tortured by perpetual pretense."

---

75   Marcus Aurelius, *Meditations,* II

Seneca, *Letters From a Stoic*[76]

"In banquets remember that you entertain two guests, body and soul: and whatever you shall have given to the body you soon eject: but what you shall have given to the soul, you keep always."

Epictetus, *Enchiridion*[77]

"Sickness is a problem for the body, not the mind — unless the mind decides that it is a problem. Lameness, too, is the body's problem, not the mind's. Say this to yourself whatever the circumstance and you will find without fail that the problem pertains to something else, not to you."

Epictetus, *Discourses*

"So what you need is not those more radical remedies which we have now finished with – blocking yourself here, being angry with yourself there, threatening yourself sternly somewhere else – but the final treatment, confidence in yourself and the belief that you are on the right path, and not led astray by the many tracks which cross yours of people who are hopeless."

---

76   Seneca, *Letters From a Stoic*, Letter 3
77   Epictetus, *Enchiridion*

Seneca, *Shortness of Life*

"Many people who have progressively lowered their personal standards in an attempt to win social acceptance and life's comforts bitterly resent those of philosophical bent who refuse to compromise their spiritual ideals and who seek to better themselves."

Epictetus, *Discourses*[78]

"If a man has reported to you, that a certain person speaks ill of you, do not make any defense to what has been told you: but reply, The man did not know the rest of my faults, for he would not have mentioned these only."

Epictetus, *Enchiridion*[79]

"He who is discontented with what he has, and with what has been granted to him by fortune, is one who is ignorant of the art of living, but he who bears that in a noble spirit, and makes reasonable use of all that comes from it, deserves to be regarded as a good man."

---

78   Epictetus, *Discourses*, II
79   Epictetus, *Enchiridion*, 32

Epictetus, *Discourses*[80]

"If someone is able to show me that what I think or do is not right, I will happily change, for I seek the truth, by which no one was ever truly harmed. It is the person who continues in his self-deception and ignorance who is harmed."

Marcus Aurelius, *Meditations*

"The object of life is not to be on the side of the majority, but to escape finding oneself in the ranks of the insane."

Marcus Aurelius, *Meditations*

"I have often wondered how it is that every man loves himself more than all the rest of men, but yet sets less value on his own opinion of himself than on the opinion of others."

Marcus Aurelius, *Meditations*

"When another blames you or hates you, or people voice similar criticisms, go to their souls, penetrate inside and see what sort of people they are. You will realize that there is no need to be racked with anxiety that they should hold any particular opinion about you."

---

[80] Epictetus, *Discourses*, II

Marcus Aurelius, *Meditations*

"Or is it your reputation that's bothering you? But look at how soon we're all forgotten. The abyss of endless time that swallows it all. The emptiness of those applauding hands. The people who praise us; how capricious they are, how arbitrary. And the tiny region it takes place. The whole earth a point in space - and most of it uninhabited."

Marcus Aurelius, *Meditations*[81]

b. **Life and Living Well** – Stoic passages on birth and the promise of the future.

"We ought to do good to others as simply as a horse runs, or a bee makes honey, or a vine bears grapes season after season without thinking of the grapes it has borne."

Marcus Aurelius, *Meditations*

"The love of power or money or luxurious living are not the only things which are guided by popular thinking. We take our cue from people's thinking even in the way we feel pain."

---

81 Marcus Aurelius, *Meditations*, II, IV, I, ibid, II

Marcus Aurelius, *Meditations*

" Our situation is like that at a festival. Sheep and cattle are driven to it to be sold, and most people come either to buy or to sell, while only a few come to look at the spectacle of the festival, to see how it is proceeding and why, and who is organizing it, and for what purpose. So also in this festival of the world. Some people are like sheep and cattle and are interested in nothing but their fodder; for in the case of those of you who are interested in nothing but your property, and land, and slaves, and public posts, all of that is nothing more than fodder. Few indeed are those who attend the fair for love of the spectacle, asking, 'What is the universe, then, and who governs it? No one at all? And yet when a city or household cannot survive for even a very short time without someone to govern it and watch over it, how could it be that such a vast and beautiful structure could be kept so well ordered by mere chance and good luck? So there must be someone governing it. What sort of being is he, and how does he govern it? And we who have been created by him, who are we, and what were we created for? Are we bound together with him in some kind of union and interrelationship, or is that not the case?' Such are the thoughts that are aroused in this small collection of people; and from then on, they devote their leisure to this one thing alone, to finding out about the festival before they have to take their leave. What comes about, then? They become an object of mockery for the crowd, just as the spectators at an ordinary festival are mocked by the traders; and even the sheep and cattle, if they had sufficient intelligence, would laugh at those who attach value to anything other than fodder!"

Epictetus, *Discourses*[82]

"Because what is a human being? Part of a community – the community of gods and men, primarily, and secondarily that of the city we happen to inhabit, which is only a microcosm of the universe in too."

Marcus Aurelius, *Meditations*[83]

"Friendship is nothing else than an accord in all things, human and divine, conjoined with mutual goodwill and affection, and I am inclined to think that, with the exception of wisdom, no better thing has been given to man by the immortal gods"

Cicero, *On Friendship*[84]

"You should, I need hardly say, live in such a way that there is nothing which you could not as easily tell your enemy as keep to yourself."

---

82   Epictetus, *Discourses*, III
83   Marcus Aurelius, *Meditations,* II, IV, XI
84   Cicero, *On Friendship*

Seneca, *Letters from a Stoic*[85]

"He who is afraid of pain will sometimes also be afraid of some of the things that will happen in the world, and even this is impiety. And he who pursues pleasure will not abstain from injustice, and this is plainly impiety."

Marcus Aurelius, *Meditations*

"Just what is the civil law? What neither influence can affect, nor power break, nor money corrupt: were it to be suppressed or even merely ignored or inadequately observed, no one would feel safe about anything, whether his own possessions, the inheritance he expects from his father, or the bequests he makes to his children."

Cicero, *Pro Gallio*[86]

"If, at some point in your life, you should come across anything better than justice, honesty, self-control, courage – than a mind satisfied that is has succeeded in enabling you to act rationally, and satisfied to accept what's beyond its control – if you find anything better than that, embrace it without reservations – it must be an extraordinary thing indeed – and enjoy it to the full."

---

85 Seneca, *Letters From a Stoic*, Letter 31
86 Cicero, *Pro Gallio*

Marcus Aurelius, *Meditations*[87]

Just as the soul fills the body, so God fills the world. Just as the soul bears the body, so God endures the world. Just as the soul sees but is not seen, so God sees but is not seen. Just as the soul feeds the body, so God gives food to the world.

Cicero, *On Old Age*[88]

"Thus Socrates became perfect, improving himself by everything. attending to nothing but reason. And though you are not yet a Socrates, you ought, however, to live as one desirous of becoming a Socrates.

Epictetus, *Discourses*[89]

"There were two vices much blacker and more serious than the rest: lack of persistence and lack of self-control… persist and resist."

---

87   Marcus Aurelius, *Meditations*, II, I, VI
88   Cicero, *On Old Age*
89   Epictetus, *Discourses*, IV

Marcus Aurelius, *Meditations*

"For while we are enclosed in these confinements of the body, we perform as a kind of duty the heavy task of necessity; for the soul from heaven has been cast down from its dwelling on high and sunk, as it were, into the earth, a place just the opposite to godlike nature and eternity. But I believe that the immortal gods have sown souls in human bodies so there might exist beings to guard the world and after contemplating the order of heaven, might imitate it by their moderation and steadfastness in life."

Cicero, *On Old Age*[90]

"What then can guide a man? One thing and only one, philosophy. But this consists in keeping the daimon within a man free from violence and unharmed, superior to pains and pleasures, doing nothing without a purpose, nor yet falsely and with hypocrisy."

Marcus, Aurelius, *Meditations*

"It is not right that anything of any other kind, such as praise from the many, or power, or enjoyment of pleasure, should come into competition with that which is rationally and politically and practically good."

---

[90] Cicero, *On Old Age*

Marcus Aurelius, *Meditations*[91]

"Yet if we place the good in right choice, the preservation of our relationships itself becomes a good. And besides, he who gives up certain external things achieves the good through that. 'My father's depriving me of money.' But he isn't causing you any harm. 'My brother is going to get the greater share of the land.' Let him have as much as he wishes. He won't be getting any of your decency, will he, or of your loyalty, or of your brotherly love? For who can disinherit you of possessions such as those? Not even Zeus; nor would he wish to, but rather he has placed all of that in my own power, even as he had it himself, free from hindrance, compulsion, and restraint."

Epictetus, *Discourses*[92]

"Just ask whether they put their self-interest in externals or in moral choice. If it's in externals, you cannot call them friends, any more than you can call them trustworthy, consistent, courageous or free."

Marcus Aurelius, *Meditations*

"My advice is really this: what we hear the philosophers saying and what we find in their writings should be applied in our pursuit of the

---

91   Marcus Aurelius, *Meditations*, II, III
92   Epictetus, *Discourses*

happy life. We should hunt out the helpful pieces of teaching, and the spirited and noble-minded sayings which are capable of immediate practical application—not far-fetched or archaic expressions or extravagant metaphors and figures of speech—and learn them so well that words become works. No one to my mind lets humanity down quite so much as those who study philosophy as if it were a sort of commercial skill and then proceed to live in a quite different manner from the way they tell other people to live."

Seneca, *Letters From a Stoic*[93]

"Everything that exists is in a manner the seed of that which will be."

Marcus Aurelius, *Meditations*

"And what else can I do, lame old man that I am, than sing the praise of God? If I were a nightingale, I would perform the work of a nightingale, and if I were a swan, that of a swan. But as it is, I am a rational being, and I must sing the praise of God.

This is my work, and I accomplish it, and I will never abandon my post for as long as it is granted to me to remain in it; and I invite all of you to join me in this same song."

---

[93] Seneca, *Letters From a Stoic*, Letter 11

Epictetus, *Discourses*[94]

"But I must at the very beginning lay down this principle—friendship can only exist between good men."

Cicero, *On Friendship*[95]

Adapt yourself to the things among which your lot has been cast and love sincerely the fellow creatures with whom destiny has ordained that you shall live.

Marcus Aurelius, *Meditations*

"Husband and wife should come together to craft a shared life, procreating children, seeing all things as shared between them-with nothing withheld or private to one another-not even their bodies. The birth of a human being which results from this union is, to be sure, something wonderful-but it isn't yet enough to account for the relationship of husband and wife-since even outside marriage it could result from any other sexual union (just as in the case of animals). So, in marriage there must be, above all, perfect companionship and mutual love - both in sickness, health and under all conditions-it

---

94   Epictetus, *Discourses, IV*
95   Cicero, *On Friendship*

should be with desire for this (and children) that both entered upon marriage."

Musonius Rufus, *How To Live*

"Life will follow the path it started upon, and will neither reverse nor check its course; it will make no noise, it will not remind you of its swiftness. Silent it will glide on; it will not prolong itself at the command of a king, or at the applause of the populace. Just as it was started on its first day, so it will run; nowhere will it turn aside, nowhere will it delay."

Seneca, *On Shortness of Life*

"Whatever is in any way beautiful hath its source of beauty in itself, and is complete in itself; praise forms no part of it. So it is none the worse nor the better for being praised."

Marcus Aurelius, *Meditations*[96]

"Suppose I should say to a wrestler, 'Show me your muscle'. And he should answer me, 'See my dumb-bells'. Your dumb-bells are your own affair; I want to see the effect of them.

---

[96]   Marcus Aurelius, *Meditations*, IV, II, ibid

"Take the treatise 'On Choice', and see how thoroughly I have perused it.

I am not asking about this, O slave, but how you act in choosing and refusing, how you manage your desires and aversions, your intentions and purposes, how you meet events -- whether you are in harmony with nature's laws or opposed to them. If in harmony, give me evidence of that, and I will say you are progressing; if the contrary, you may go your way, and not only comment on your books, but write some like them yourself; and what good will it do you?"

Epictetus, *On Choice*[97]

"Indeed the state of all who are preoccupied is wretched, but the most wretched are those who are toiling not even at their own preoccupations, but must regulate their sleep by another's, and their walk by another's pace, and obey orders in those freest of all things, loving and hating. If such people want to know how short their lives are, let them reflect how small a portion is their own."

---

97   Epictetus, *On Choice*

Seneca, *On Shortness of Life*[98]

"What is your art? To be good. And how is this accomplished well except by general principles, some about the nature of the universe, and others about the proper constitution of man?"

Marcus Aurelius, *Meditations*

"Since it's clear then that what sets itself in motion is eternal, who could fail to attribute such a nature to the soul. Anything set in motion by external impetus is inanimate; what is animate moves by its own interior impulse. This is the nature and power of soul. And because it is the one thing out of all that sets itself in motion, then surely it was never born and will last forever."

Cicero, *On Living and Dying Well*[9]

"But when you are looking on anyone as a friend when you do not trust him as you trust yourself, you are making a grave mistake, and have failed to grasp sufficiently the full force of true friendship."

---

98   Seneca, *On Shortness of Life*
99   Cicero, *On Living and Dying Well*

Seneca, *Letters From a Stoic*[100]

"We must stand up against old age and make up for its drawbacks by taking pains. We must fight it as we should an illness. We must look after our health, use moderate exercise, take just enough food and drink to recruit, but not to overload, our strength. Nor is it the body alone that must be supported, but the intellect and soul much more."

Cicero, *On Old Age*[101]

"And in the case of superior things like stars, we discover a kind of unity in separation. The higher we rise on the scale of being, the easier it is to discern a connection even among things separated by vast distances."

Marcus Aurelius, *Meditations*

"These reasonings are unconnected: 'I am richer than you, therefore I am better'; 'I am more eloquent than you, therefore I am better.' The connection is rather this: 'I am richer than you, therefore my property is greater than yours;' 'I am more eloquent than you, therefore my style is better than yours.' But you, after all, are neither property nor style."

---

100   Seneca, *Letters From a Stoic*, Letter 36
101   Cicero, *On Old Age*

Epictetus, *Enchiridion*

"It is better to do wrong seldom and to own it, and to act right for the most part, than seldom to admit that you have done wrong and to do wrong often."

Epictetus, Enchiridion[102]

"How can life be worth living, to use the words of Ennius, which lacks that repose which is to be found in the mutual good-will of a friend? What can be more delightful than to have some one to whom you can say everything with the same absolute confidence as to yourself? Is not prosperity robbed of half its value if you have no one to share your joy? On the other hand, misfortunes would be hard to bear if there were not some one to feel them even more acutely than yourself."

Cicero, *On Friendship*[103]

"Very little is needed to make a happy life; it is all within yourself in your way of thinking."

---

102   Epictetus, *Enchiridion*, 18
103   Cicero, *On Friendship*

Marcus Aurelius, *Meditations*[104]

"If you wish your house to be well managed, imitate the Spartan Lycurgus. For as he did not fence his city with walls, but fortified the inhabitants by virtue and preserved the city always free; so do you not cast around (your house) a large court and raise high towers, but strengthen the dwellers by good-will and fidelity and friendship, and then nothing harmful will enter it, not even if the whole band of wickedness shall array itself against it.

Epictetus, *Enchiridion*[105]

"I conclude, then, that the plea of having acted in the interests of a friend is not a valid excuse for a wrong action. . . . We may then lay down this rule of friendship--neither ask nor consent to do what is wrong. For the plea "for friendship's sake" is a discreditable one, and not to be admitted for a moment."

Cicero, *On Living Well*[106]

"As for myself, I can only exhort you to look on Friendship as the most valuable of all human possessions, no other being equally suited to the moral nature of man, or so applicable to every state and

---

104   Marcus Aurelius, *Meditations*, IV, III
105   Epictetus, *Enchiridion*, 20
106   Cicero, *On Living Well*

circumstance, whether of prosperity or adversity, in which he can possibly be placed. But at the same time I lay it down as a fundamental axiom that "true Friendship can only subsist between those who are animated by the strictest principles of honour and virtue." When I say this, I would not be thought to adopt the sentiments of those speculative moralists who pretend that no man can justly be deemed virtuous who is not arrived at that state of absolute perfection which constitutes, according to their ideas, the character of genuine wisdom. This opinion may appear true, perhaps, in theory, but is altogether inapplicable to any useful purpose of society, as it supposes a degree of virtue to which no mortal was ever capable of rising."

Cicero, *Letters*[107]

"We are at the mercy of whoever wields authority over the things we either desire or detest. If you would be free, then, do not wish to have, or avoid, things that other people control, because then you must serve as their slave."

Epictetus, *Discourses*[108]

"If it is not right do not do it; if it is not true do not say it."

---

107   Cicero, *Letters*
108   Epictetus, *Discourses*, III

Marcus Aurelius, *Meditations*[109]

c. **What to do With Work** – Advice concerning heavy workloads and positive reinforcement.

"If you seek tranquility, do less. Or do what's essential – what the logos of a social being requires, and in the requisite way. Which brings a double satisfaction: to do less, better. Because most of what we say and do is not essential. If you can eliminate it, you'll have more time, and more tranquility. Ask yourself at every moment, 'Is this necessary?'"

Marcus Aurelius, *Meditations*

"For in this Case, we are not to give Credit to the Many, who say, that none ought to be educated but the Free; but rather to the Philosophers, who say, that the Well-educated alone are free."

Epictetus, *Fragments*[110]

"Think of your many years of procrastination; how the gods have repeatedly granted you further periods of grace, of which you have taken no advantage. It is time now to realize the nature of the universe to which you belong, and of that controlling Power whose offspring

---

109　　Marcus Aurelius, *Meditations*, II, ibid
110　　Epictetus, Fragment

you are; and to understand that your time has a limit set to it. Use it, then, to advance your enlightenment; or it will be gone, and never in your power again."

Marcus Aurelius, *Meditations*

A key point to bear in mind: The value of attentiveness varies in proportion to its object. You're better off not giving the small things more time than they deserve.

Marcus Aurelius, *Meditations*

"You should rather suppose that those are involved in worthwhile duties who wish to have daily as their closest friends Zeno, Pythagoras, Democritus and all the other high priests of science, and Aristotle and Theophrastus. None of these will be too busy to see you, none of these will not send his visitor away happier and more devoted to himself, none of these will allow anyone to depart empty-handed. They are at home to all mortals by night and by day."

Seneca, *On Shortness of Life*[111]

We should not be so taken up in the search for truth, as to neglect the needful duties of active life; for it is only action that gives a true value and commendation to virtue.

Cicero, *On Old Age*[112]

"You need to avoid certain things in your train of thought: everything random, everything irrelevant. And certainly everything self-important or malicious. You need to get used to winnowing your thoughts, so that if someone says, "What are your thinking about?" you can respond at once (and truthfully) that you are thinking this or thinking that."

Marcus Aurelius, *Meditations*[113]

There is nothing that we can properly call our own but our time, and yet everybody fools us out of it who has a mind to do it. If a man borrows a paltry sum of money, there must needs be bonds and securities, and every common civility is presently charged upon account. But he who has my time thinks he owes me nothing for it, though it be a debt that gratitude itself can never repay.

---

111   Seneca, *On Shortness of Life*
112   Cicero, *On Old Age*
113   Marcus Aurelius, *Meditations*, II, XI, III

Seneca, *Letters From a Stoic*[114]

"I have the better right to indulge herein, because my devotion to letters strengthens my oratorical powers, and these, such as they are, have never failed my friends in their hour of peril. Yet insignificant though these powers may seem to be, I fully realize from what source I draw all that is highest in them. Had I not persuaded myself from my youth up, thanks to the moral lessons derived from a wide reading, that nothing is to be greatly sought after in this life save glory and honor, and that in their quest all bodily pains and all dangers of death or exile should be lightly accounted, I should never have borne for the safety of you all the burnt of many a bitter encounter, or bared my breast to the daily onsets of abandoned persons. All literature, all philosophy, all history, abounds with incentives to noble action, incentives which would be buried in black darkness were the light of the written word not flashed upon them."

Cicero, *On Old Age*[115]

"To be sure, external things of whatever kind require skill in their use, but we must not grow attached to them; whatever they are, they should only serve for us to show how skilled we are in our handling of them."

---

114   Seneca, *Letters From a Stoic*, Letter 53
115   Cicero, *On Old Age*

Marcus Aurelius, *Meditations*

"If you have been placed in a position above others, are you automatically going to behave like a despot? Remember who you are and whom you govern – that they are kinsmen, brothers by nature, fellow descendants of Zeus."

Marcus Aurelius, *Meditations*[116]

"If you apply yourself to study you will avoid all boredom with life, you will not long for night because you are sick of daylight, you will be neither a burden to yourself nor useless to others, you will attract many to become your friends and the finest people will flock about you."

Seneca, *Letters From a Stoic*[117]

"Don't just say you have read books. Show that through them you have learned to think better, to be a more discriminating and reflective person. Books are the training weights of the mind. They are very helpful, but it would be a bad mistake to suppose that one has made progress simply by having internalized their contents."

---

116   Marcus Aurelius, *Meditations*, V, II, IV
117   Seneca, *Letters From a Stoic*, Letter 21

Epictetus, *Discourses*[118]

'Show me one person who cares how they act, someone for whom success is less important than the manner in which it is achieved. While out walking, who gives any thought to the act of walking itself? Who pays attention to the process of planning, not just the outcome?"

Marcus Aurelius, *Meditations*

"Hour by hour resolve firmly to do what comes to hand with dignity, and with humanity, independence, and justice. Allow your mind freedom from all other considerations. This you can do, if you will approach each action as though it were your last, dismissing the desire to create an impression, the admiration of self, the discontent with your lot. See how little man needs to master, for his days to flow on in quietness and piety: he has but to observe these few counsels, and the gods will ask nothing more."

Marcus Aurelius, *Meditations*[119]

"Though, even if there were no such great advantage to be reaped from it, and if it were only pleasure that is sought from these studies, still I imagine you would consider it a most reasonable and liberal employment of the mind: for other occupations are not suited to every time, nor to every age or place; but these studies are the food of youth,

---

118   Epictetus, *Discourses*, II
119   Marcus Aurelius, *Meditations*, II

the delight of old age; the ornament of prosperity, the refuge and comfort of adversity; a delight at home, and no hindrance abroad; they are companions by night, and in travel, and in the country."

Cicero, *On Oration*[120]

"To want to know more than is sufficient is a form of intemperance. Apart from which this kind of obsession with the arts turns people into pedantic, irritating, tactless, self-satisfied bores, not learning what they need simply because they spend their time learning things they will never need. The scholar Didymus wrote four thousand works: I should feel sorry him if he had merely read so many useless works."

Seneca, *Letters From a Stoic*

"Tentative efforts lead to tentative outcomes. Therefore, give yourself fully to your endeavors. Decide to construct your character through excellent actions and determine to pay the price of a worthy goal. The trials you encounter will introduce you to your strengths. Remain steadfast...and one day you will build something that endures: something worthy of your potential."

---

120   Cicero, *On Oration*

Epictetus, Discourses[121]

"Why do you want to read anyway – for the sake of amusement or mere erudition? Those are poor, fatuous pretexts. Reading should serve the goal of attaining peace; if it doesn't make you peaceful, what good is it?"

Epictetus, *Of Human Freedom*

"But only philosophy will wake us; only philosophy will shake us out of that heavy sleep. Devote yourself entirely to her. You're worthy of her, she's worthy of you-fall into each other's arms. Say a firm, plain no to every other occupation."

Seneca, *Letters From a Stoic*[122]

"If, on the other hand, we read books entitled On Impulse not just out of idle curiosity, but in order to exercise impulse correctly; books entitled On Desire and On Aversion so as not to fail to get what we desire or fall victim to what we would rather avoid; and books entitled On Moral Obligation in order to honor our relationships and never do anything that clashes or conflicts with this principle; then we wouldn't get frustrated and grow impatient with our reading.

121　Epictetus, *Discourses, IV*
122　Seneca, *Letters From a Stoic*, Letter 40, Letter 8

Instead we would be satisfied to act accordingly. And rather than reckon, as we are used to doing, 'How many lines I read, or wrote, today,' we would pass in review how 'I applied impulse today the way the philosophers recommend"

Epictetus, *Of Human Freedom*

"We ought, then, to set up images of a kind that can adhere longest in the memory. And we shall do so if we establish likenesses as striking as possible; if we set up images that are not many or vague, but doing something; if we assign to them exceptional beauty or singular ugliness; if we dress some of them with crowns or purple cloaks, for example, so that the likeness may be more distinct to us; or if we somehow disfigure them, as by introducing one stained with blood or soiled with mud or smeared with red paint, so that its form is more striking, or by assigning certain comic effects to our images, for that, too, will ensure our remembering them more readily."

Cicero, *Letters*[123]

"In literature, too, it is not great achievement to memorize what you have read while not formulating an opinion of your own."

---

123   Cicero, *Letters*

Epictetus, *Discourses*[124]

"Finally, everybody agrees that no one pursuit can be successfully followed by a man who is preoccupied with many things—eloquence cannot, nor the liberal studies—since the mind, when distracted, takes in nothing very deeply, but rejects everything that is, as it were, crammed into it. There is nothing the busy man is less busied with than living: there is nothing that is harder to learn."

Seneca, *Letters From a Stoic*[125]

"Of this last kind of comparisons is that quoted from the elder Cato, who, when asked what was the most profitable thing to be done on an estate, replied, "To feed cattle well." "What second best?" "To feed cattle moderately well." "What third best?" "To feed cattle, though but poorly." "What fourth best?" "To plough the land." And when he who had made these inquiries asked, "What is to be said of making profit by usury?" Cato replied, "What is to be said of making profit by murder?""

Cicero, *On Duties*[126]

d. **Open to Possibilities** – What the Stoics thought about views and anticipation of the future.

---

124   Epictetus, *Discourses*, IV
125   Seneca, *Letters From a Stoic*, Letter 13
126   Cicero, *On Duties*

"My thought for today is something which I found in Epicurus (yes, I actually make a practice of going over to the enemy's camp – by way of reconnaissance, not as a deserter!). 'A cheerful poverty,' he says, 'is an honorable state.' But if it is cheerful it is not poverty at all. It is not the man who has too little who is poor, but the one who hankers after more. What difference does it make how much there is laid away in a man's safe or in his barns, how many head of stock he grazes or how much capital he puts out at interest, if he is always after what is another's and only counts what he has yet to get, never what he has already. You ask what is the proper limit to a person's wealth? First, having what is essential, and second, having what is enough."

Seneca, *Letters From a Stoic*[127]

"People who are physically ill are unhappy with a doctor who doesn't give them advice, because they think he has given up on them. Shouldn't we feel the same towards a philosopher – and assume that he has given up hope of our ever becoming rational – if he will no longer tell us what we need (but may not like) to hear?"

---

127   Seneca, *Letters From a Stoic*, Letter 11

Marcus Aurelius, *Meditations*[128]

"The best Armour of Old Age is a well spent life preceding it; a Life employed in the Pursuit of useful Knowledge, in honourable Actions and the Practice of Virtue; in which he who labours to improve himself from his Youth, will in Age reap the happiest Fruits of them; not only because these never leave a Man, not even in the extremest Old Age; but because a Conscience bearing Witness that our Life was well-spent, together with the Remembrance of past good Actions, yields an unspeakable Comfort to the Soul"

Cicero, *Letters*[129]

"So when you hear that even life and the like are indifferent, don't become apathetic; and by the same token, when you're advised to care about them, don't become superficial and conceive a passion for externals."

Marcus Aurelius, *Meditation*

"Remember that the divine order is intelligent and fundamentally good. Life is not a series of random, meaningless episodes, but an ordered, elegant whole that follows ultimately comprehensible laws."

---

128   Marcus Aurelius, *Meditations*, I, X, IV
129   Cicero, *Letters*

Epictetus, *Discourses*[130]

"Pain too is just a scary mask: look under it and you will see. The body sometimes suffers, but relief is never far behind. And if that isn't good enough for you, the door stands open; otherwise put up with it. The door needs to stay open whatever the circumstances, with the result that our problems disappear."

Marcus Aurelius, *Meditations*

"Do not be whirled about, but in every movement have respect to justice, and on the occasion of every impression maintain the faculty of comprehension or understanding."

Marcus Aurelius, *Meditations*

"From the philosopher Catulus, never to be dismissive of a friend's accusation, even if it seems unreasonable, but to make every effort to restore the relationship to its normal condition."

---

130   Epictetus, *Discourses*

Marcus Aurelius, *Meditations*

"Set aside a certain number of days, during which you shall be content with the scantiest and cheapest fare, with course and rough dress, saying to yourself the while: 'Is this the condition that I feared?'"

Seneca, *Letters From a Stoic*

"This presumption that you possess knowledge of any use has to be dropped before you approach philosophy – just as if we were enrolling in a school of music or mathematics."

Marcus Aurelius, *Meditations*

"There are two things that must be rooted out in human beings - arrogant opinion and mistrust. Arrogant opinion expects that there is nothing further needed, and mistrust assumes that under the torrent of circumstance there can be no happiness."

Epictetus, *Discourses*[131]

"Once you have rid yourself of the affliction there, though, every change of scene will become a pleasure. You may be banished to the ends of the earth, and yet in whatever outlandish corner of the world

---

131   Epictetus, *Discourses, IV*

you may find yourself stationed, you will find that place, whatever it may be like, a hospitable home. Where you arrive does not matter so much as what sort of person you are when you arrive there."

Seneca, *Letters From a Stoic*[132]

"You might as well get on your knees and pray that your nose won't run. A better idea would be to wipe your nose and forgo the prayer. The point is, isn't there anything God gave you for your present problem?"

Marcus Aurelius, *Meditations*[133]

"Why are we not angry if we are told that we have a headache, and why are we angry if we are told that we reason badly, or choose wrongly?" The reason is that we are quite certain that we have not a headache, or are not lame, but we are not so sure that we make a true choice. So having assurance only because we see with our whole sight, it puts us into suspense and surprise when another with his whole sight sees the opposite, and still more so when a thousand others deride our choice. For we must prefer our own lights to those of so many others, and that is bold and difficult."

---

132   Seneca, *Letters From a Stoic*, Letter 35, Letter 6
133   Marcus Aurelius, *Meditations*, IV, X, ibid, V

Epictetus, *Discourses*

"The day has already begun to lessen. It has shrunk considerably, but yet will still allow a goodly space of time if one rises, so to speak, with the day itself. We are more industrious, and we are better men if we anticipate the day and welcome the dawn;"

Seneca, *Letters From a Stoic*

"Now there are two kinds of hardening, one of the understanding, the other of the sense of shame, when a man is resolved not to assent to what is manifest nor to desist from contradictions. Most of us are afraid of mortification of the body, and would contrive all means to avoid such a thing, but we care not about the soul's mortification. And indeed with regard to the soul, if a man be in such a state as not to apprehend anything, or understand at all, we think that he is in a bad condition; but if the sense of shame and modesty are deadened, this we call even power (or strength)."

Epictetus, *Discourses*[134]

"When, therefore, you see anyone eminent in honors, or power, or in high esteem on any other account, take heed not to be hurried away with the appearance, and to pronounce him happy; for, if the essence

---

134   Epictetus, *Discourses*, IV, ibid

of good consists in things in our own control, there will be no room for envy or emulation. But, for your part, don't wish to be a general, or a senator, or a consul, but to be free;"

Epictetus, *Enchiridion*[135]

---

[135] Epictetus, *Enchiridion*, 3

III. BEAT THE HEAT WITH SUMMER VIRTUE

a. **Relaxing: Not Just for Kids** – The Stoics on leisure.

"The man who spends his time choosing one resort after another in a hunt for peace and quiet, will in every place he visits find something to prevent him from relaxing."

Marcus Aurelius, *Meditations*[136]

"Nothing, to my way of thinking, is a better proof of a well ordered mind than a man's ability to stop just where he is and pass some time in his own company."

Seneca, *Letters From a Stoic*[137]

"These studies which stimulate the young, divert the old, are an ornament in prosperity and a refuge and comfort in adversity; they delight us at home, are no impediment in public life, keep us company at night, in our travels, and whenever we retire to the country."

---

136   Marcus Aurelius, *Meditations*, V
137   Seneca, *Letters From a Stoic*, Letter 30, Letter 26

Cicero, *Pro Archia*

"At times we ought to drink even to intoxication, not so as to drown, but merely to dip ourselves in wine, for wine washes away troubles and dislodges them from the depths of the mind and acts as a remedy to sorrow as it does to some diseases. The inventor of wine is called Liber, not from the license which he gives to our tongues but because he liberates the mind from the bondage of cares and emancipates it, animates it and renders it more daring in all that it attempts."

Seneca, *Letters From a Stoic*

"If you ever happen to turn your attention to externals, for the pleasure of any one, be assured that you have ruined your scheme of life. Be contented, then, in everything, with being a philosopher; and if you with to seem so likewise to any one, appear so to yourself, and it will suffice you."

Epictetus, *Enchiridion*[138]

"For other forms of relaxation are not so universally suited to all ages, times, and places; but these studies [of literature] sustain youth and entertain old age, they enhance prosperity, and offer a refuge and solace in adversity, they delight us when we are at home without

---

138   Epictetus, *Enchiridion*, 14

hindering us in the wider world, and are with us at night, when we travel and when we visit the countryside."

Cicero, *Pro Archia*[139]

"Cling, therefore, to this sound and wholesome plan of life; indulge the body just so far as suffices for good health. ... Your food should appease your hunger, your drink quench your thirst, your clothing keep out the cold, your house be a protection against inclement weather. It makes no difference whether it is built of turf or variegated marble imported from another country: what you have to understand is that thatch makes a person just as good a roof as gold."

Seneca, *Letters From a Stoic*[140]

"We must consider what is the time for singing, what the time for play, and in whose presence: what will be unsuited to the occasion; whether our companions are to despise us, or we to despise ourselves: when to jest, and whom to mock at: and on what occasion to be conciliatory and to whom: in a word, how one ought to maintain one's character in society. Wherever you swerve from any of these principles, you suffer loss at once; not loss from without, but issuing from the very act itself."

---

139   Cicero, *Pro Archia*
140   Seneca, *Letters From a Stoic*, Letter 21, Letter 35

Epictetus, *Discourses*[141]

"We must go for walks out of doors, so that the mind can be strengthened and invigorated by a clear sky and plenty of fresh air. At times it will acquire fresh energy from a journey by carriage and a change of scene, or from socializing and drinking freely. Occasionally we should even come to the point of intoxication, sinking into drink but not being totally flooded by it; for it does wash away cares, and stirs the mind to its depths, and heals sorrow just as it heals certain diseases."

Seneca, *Letters From a Stoic*

"Let us assume that entertainment is the sole end of reading; even so I think you would hold that no mental employment is so broadening to the sympathies or so enlightening to the understanding. Other pursuits belong not to all times, all ages, all conditions; but this gives stimulus to our youth and diversion to our old age; this adds a charm to success, and offers a haven of consolation to failure. Through the night-watches, on all our journeys, and in our hours of ease, it is our unfailing companion."

---

141   Epictetus, *Discourses*, III

Cicero, *Letters*[142]

"True happiness is to enjoy the present, without anxious dependence upon the future, not to amuse ourselves with either hopes or fears but to rest satisfied with what we have, which is sufficient, for he that is so wants nothing. The greatest blessings of mankind are within us and within our reach. A wise man is content with his lot, whatever it may be, without wishing for what he has not."

Seneca, *Letters From a Stoic*

b. **On Temperament and Temperature** – Stoic advice on environmental and physical stress.

"Indulge the body just so far as suffices for good health. It needs to be treated somewhat strictly to prevent it from being disobedient to the spirit. Your food should appease your hunger, your drink quench your thirst, your clothing keep out the cold, your house be a protection against inclement weather."

---

142   Cicero, *Letters*

Seneca, *Letters From a Stoic*[143]

"Objective judgment, now, at this very moment. Unselfish action, now, at this very moment. Willing acceptance—now, at this very moment—of all external events. That's all you need."

Marcus Aurelius, *Meditations*[144]

"Everywhere means nowhere. When a person spends all his time in foreign travel, he ends by having many acquaintances, but no friends. And the same thing must hold true of men who seek intimate acquaintance with no single author, but visit them all in a hasty and hurried manner. 3. Food does no good and is not assimilated into the body if it leaves the stomach as soon as it is eaten; nothing hinders a cure so much as frequent change of medicine; no wound will heal when one salve is tried after another; a plant which is often moved can never grow strong. There is nothing so efficacious that it can be helpful while it is being shifted about. And in reading of many books is distraction."

---

143 Seneca, *Letters From a Stoic*, Letter 18, Letter 6
144 Marcus Aurelius, *Meditations*, X

Seneca, *Shortness of Life*

"What good are gilded rooms or precious stones-fitted on the floor, inlaid in the walls, carried from great distances at the greatest expense? These things are pointless and unnecessary-without them isn't it possible to live healthy? Aren't they the source of constant trouble? Don't they cost vast sums of money that, through public and private charity, may have benefited many?"

Musonius Rufus, *How To Live*[145]

"Is this all the habit you acquired when you studied philosophy, to look to others and to hope for nothing from yourself and your own acts? Lament therefore and mourn, and when you eat be fearful that you will have nothing to eat to-morrow. Tremble for your wretched slaves, lest they should steal, or run away, or die. Live in this spirit, and never cease to live so, you who never came near philosophy, except in name, and disgraced its principles so far as in you lies, by showing them to be useless and unprofitable to those who take them up."

---

145   Musonius Rufus, *How To Live*

Epictetus, *Discourses*[146]

"As far as I am concerned, I know that I have lost not wealth but distractions. The body's needs are few: it wants to be free from cold, to banish hunger and thirst with nourishment; if we long for anything more we are exerting ourselves to serve our vices, not our needs."

Seneca, *On Shortness of Life*[147]

"In a sense, people are our proper occupation. Our job is to do them good and put up with them.

But when they obstruct our proper tasks, they become irrelevant to us—like sun, wind, animals. Our actions may be impeded by them, but there can be no impeding our intentionsor our dispositions. Because we can accommodate and adapt. The mind adapts and converts to its own purposes the obstacle to our acting. The impediment to action advances action. What stands in the way becomes the way."

Marcus Aurelius, *Meditations*

"What difference does it make how much is laid away in a man's safe or in his barns, how many head of stock he grazes or how much capital he puts out at interest, if he is always after what is another's and only

---

146   Epictetus, *Discourses*, IV
147   Seneca, *On Shortness of Life*, II

counts what he has yet to get, never what he has already? You ask what is the proper limit to a person's wealth? First, having what is essential, and second, having what is enough."

Seneca, *Letters From a Stoic*[148]

"First to those universal principles I have spoken of: these you must keep at command, and without them neither sleep nor rise, drink nor eat nor deal with men: the principle that no one can control another's will, and that the will alone is the sphere of good and evil."

Epictetus, *Discourses*[149]

"Everyone has the obligation to ponder well his own specific traits of character. He must also regulate them adequately and not wonder whether someone else's traits might suit him better. The more definitely his own a man's character is, the better it fits him."

Cicero, *Letters*[150]

"Just as plants receive nourishment for survival, not pleasure-for humans, food is the medicine of life. Therefore it is appropriate for us

---

148 Seneca, *Letters From a Stoic*, Letter 28
149 Epictetus, *Discourses*, II
150 Cicero, *Letters*

to eat for living, not pleasure, especially if we want to follow the wise words of Socrates, who said most men live to eat: I eat to live"

Musonius Rufus, *How To Live*[151]

"I am not a 'wise man,' nor . . . shall I ever be. And so require not from me that I should be equal to the best, but that I should be better than the wicked. It is enough for me if every day I reduce the number of my vices, and blame my mistakes."

Seneca, *Letters From a Stoic*[152]

"Retire into thyself. The rational principle which rules has this nature, that it is content with itself when it does what is just, and so secures tranquility."

Marcus Aurelius, *Meditations*[153]

"Philosophy calls for simple living, not for doing penance, and the simple way of life need not be a crude one."

---

151   Musonius Rufus, *How To Live*
152   Seneca, *Letters From a Stoic*, Letter 2, Letter 13
153   Marcus Aurelius, *Meditations*, VIII

Seneca, *Letters From a Stoic*

"For as I like a man in whom there is something of the old, so I like a man in whom there is something of the young; and he who follows this maxim, in body will possibly be an old man but he will never be an old man in mind."

Cicero, *On Old Age*[154]

"Don't let your imagination be crushed by life as a whole. Don't try to picture everything bad that could possibly happen. Stick with the situation at hand, and ask, "Why is this so unbearable? Why can't I endure it?" You'll be embarrassed to answer. Then remind yourself that past and future have no power over you. Only the present—and even that can be minimized. Just mark off its limits.

Marcus Aurelius, *Meditations*

"If a person gave your body to any stranger he met on is way, you would certainly be angry. And do you feel no shame in handing over your own mind to be confused and mystified by anyone who happens to verbally attack you?"

---

154   Cicero, *On Old Age*

Epictetus, *Enchiridion*[155]

"What progress, you ask, have I made? I have begun to be a friend to myself." That was indeed a great benefit; such a person can never be alone. You may be sure that such a man is a friend to all mankind."

Seneca, *Letters From a Stoic*

"If you live in harmony with nature you will never be poor; if you live according what others think, you will never be rich."

Seneca, *Letters From a Stoic*

"Not to waste time on nonsense. Not to be taken in by conjurors and hoodoo artists with their talk about incantations and exorcism and all the rest of it. Not to be obsessed with quail-fighting or other crazes like that."

Marcus Aurelius, *Meditations*[156]

"All outdoors may be bedlam, provided there is no disturbance within."

---

155 Epictetus, *Enchiridion*
156 Marcus Aurelius, *Meditations*, VII

Seneca, *Letters From a Stoic*[157]

"But nothing will help quite so much as just keeping quiet, talking with other people as little as possible, with yourself as much as possible. For conversation has a kind of charm about it, an insinuating and insidious something that elicits secrets from us just like love or liquor. Nobody will keep the things he hears to himself, and nobody will repeat just what he hears and no more. Neither will anyone who has failed to keep a story to himself keep the name of his informant to himself. Every person without exception has someone to whom he confides everything that is confided to himself. Even supposing he puts some guard in his garrulous tongue and is content with a single pair of ears, he will still be the creator of a host of later listeners – such is the way in which what was but a little while before a secret becomes common rumor."

Seneca, *Letters From a Stoic*

"Kindness is unconquerable, so long as it is without flattery or hypocrisy. For what can the most insolent man do to you, if you contrive to be kind to him, and if you have the chance gently advise and calmly show him what is right...and point this out tactfully and

---

157   Seneca, *Letters From a Stoic*, Letter 22, Letter 10, Letter 31, Letter 3

from a universal perspective. But you must not do this with sarcasm or reproach, but lovingly and without anger in your soul."

Marcus Aurelius, *Meditations*[158]

"In truth, Serenus, I have for a long time been silently asking myself to what I should liken such a condition of mind, and I can find nothing that so closely approaches it as the state of those who, after being released from a long and serious illness, are sometimes touched with fits of fever and slight disorders, and, freed from the last traces of them, are nevertheless disquieted with mistrust, and, though now quite well, stretch out their wrist to a physician and complain unjustly of any trace of heat in their body. It is not, Serenus, that these are not quite well in body, but that they are not quite used to being well; just as even a tranquil sea will show some ripple, particularly when it has just subsided after a storm. What you need, therefore, is not any of those harsher measures which we have already left behind, the necessity of opposing yourself at this point, of being angry with yourself at that, of sternly urging yourself on at another, but that which comes last - confidence in yourself and the belief that you are on the right path, and have not been led astray by the many cross- tracks of those who are roaming in every direction, some of whom are wandering very near the path itself. But what you desire is something great and supreme and very near to being a god - to be unshaken. "

---

158   Marcus Aurelius, *Meditations*, IV, V

Seneca, *Letters From a Stoic*

"Asia and Europe: tiny corners of the Cosmos. Every sea: a mere drop. Mount Athos: a lump of dirt. The present moment is the smallest point in all eternity. All is microscopic, changeable, disappearing. All things come from that faraway place, either originating directly from that governing part which is common to all, or else following from it as consequences. So even the gaping jaws of the lion, deadly poison, and all harmful things like thorns or an oozing bog are products of that awesome and noble source. Do not imagine these things to be alien to that which you revere, but turn your Reason to the source of all things."

Marcus Aurelius, *Meditations*

"It does good also to take walks out of doors, that our spirits may be raised and refreshed by the open air and fresh breeze: sometimes we gain strength by driving in a carriage, by travel, by change of air, or by social meals and a more generous allowance of wine."

Seneca, *Letters From a Stoic*[159]

"When you run up against someone else's shamelessness, ask yourself this: Is a world without shamelessness possible?

---

[159] Seneca, *Letters From a Stoic*, Letter 22, Letter 16

No. Then don't ask the impossible. There have to be shameless people in the world. This is one of them. The same for someone vicious or untrustworthy, or with any other defect. Remembering that the whole world class has to exist will make you more tolerant of its members."

Marcus Aurelius, *Meditations*

"Are you not scorched by the heat? Are you not cramped for room? Have you not to bathe with discomfort? Are you not drenched when it rains? Have you not to endure the clamor and shouting and such annoyances as these? Well, I suppose you set all this over against the splendour of the spectacle and bear it patiently. What then? have you not received greatness of heart, received courage, received fortitude? What care I, if I am great of heart, for aught that can come to pass? What shall cast me down or disturb me? What shall seem painful? Shall I not use the power to the end for which I received it, instead of moaning and wailing over what comes to pass?"

Epictetus, *Discourses*[160]

"Which is recorded of Socrates, that he was able both to abstain from, and to enjoy, those things which many are too weak to abstain from,

---

[160] Epictetus, *Discourses*, III

and cannot enjoy without excess. But to be strong enough both to bear the one and to be sober in the other is the mark of a man who has a perfect and invincible soul."

Marcus Aurelius, *Meditations*

"All things of the body stream away like a river, all things of the mind are dreams and delusion; life is warfare, and a visit to a strange land; the only lasting fame is oblivion."

Marcus Aurelius, *Meditations*[161]

"Two distinctive traits especially identify beyond a doubt a strong and dominant character. One trait is contempt for external circumstances, when one is convinced that men ought to respect, to desire, and to pursue only what is moral and right, that men should be subject to nothing, not to another man, not to some disturbing passion, not to Fortune.

The second trait, when your character has the disposition I outlined just now, is to perform the kind of services that are significant and most beneficial; but they should also be services that are a severe challenge, that are filled with ordeals, and that endanger not only your life but also the many comforts that make life attractive.

---

161   Marcus Aurelius, *Meditations*, XI, V, II

Of these two traits, all the glory, magnificence, and the advantage, too, let us not forget, are in the second, while the drive and the discipline that make men great are in the former."

Cicero, *De Officiis*[162]

"Is your cucumber bitter? Throw it away. Are there briars in your path? Turn aside. That is enough. Do not go on and say, "Why were things of this sort ever brought into this world?" neither intolerable nor everlasting - if thou bearest in mind that it has its limits, and if thou addest nothing to it in imagination. Pain is either an evil to the body (then let the body say what it thinks of it!)-or to the soul. But it is in the power of the soul to maintain its own serenity and tranquility. . . ."

Marcus Aurelius, *Meditations*[163]

c. **Living in Your** Prime – How the Stoics viewed those in their prime and what to do with their vigor.

"Remember to act always as if you were at a symposium. When the food or drink comes around, reach out and take some politely; if it passes you by don't try pulling it back. And if it has not reached you yet, don't let your desire run ahead of you, be patient until your

---

162   Cicero, *De Officiis*

163   Marcus Aurelius, *Meditations*, II, I, IX

turn comes. Adopt a similar attitude with regard to children, wife, wealth and status, and in time, you will be entitled to dine with the gods. Go further and decline these goods even when they are on offer and you will have a share in the gods' power as well as their company. That is how Diogenes, Heraclitus and philosophers like them came to be called, and considered, divine."

Epictetus, *Enchiridion*

"Withdraw into yourself, as far as you can. Associate with those who will make a better man of you. Welcome those whom you yourself can improve. The process is mutual; for men learn while they teach."

Seneca, *Letters From a Stoic*[164]

"The first and most important field of philosophy is the application of principles such as "Do not lie." Next come the proofs, such as why we should not lie. The third field supports and articulates the proofs, by asking, for example, "How does this prove it? What exactly is a proof, what is logical inference, what is contradiction, what is truth, what is falsehood?" Thus, the third field is necessary because of the second, and the second because of the first. The most important, though, the one that should occupy most of our time, is the first. But we do just

---

164   Seneca, *Letters From a Stoic*, Letter 24

the opposite. We are preoccupied with the third field and give that all our attention, passing the first by altogether. The result is that we lie – but have no difficulty proving why we shouldn't."

Epictetus, *Enchiridion*[165]

"Reflect on the other social roles you play. If you are a council member, consider what a council member should do. If you are young, what does being young mean, if you are old, what does age imply, if you are a father, what does fatherhood entail? Each of our titles, when reflected upon, suggests the acts appropriate to it."

Marcus Aurelius, *Meditations*

"The name of peace is sweet, and the thing itself is beneficial, but there is a great difference between peace and servitude. Peace is freedom in tranquility, servitude is the worst of all evils, to be resisted not only by war, but even by death."

---

[165] Epictetus, *Enchiridion*

Cicero, *De Oratore*[166]

"It is not the man who has too little who is poor, but the one who hankers after more."

Seneca, *Letters From a Stoic*[167]

"If all emotions are common coin, then what is unique to the good man?

To welcome with affection what is sent by fate. Not to stain or disturb the spirit within him with a mess of false beliefs. Instead, to preserve it faithfully, by calmly obeying God – saying nothing untrue, doing nothing unjust. And if the others don't acknowledge it – this life lived in   simplicity, humility, cheerfulness – he doesn't resent them for it, and isn't deterred from following the road where it leads: to the end of life. An end to be approached in purity, in serenity, in acceptance, in peaceful unity with what must be."

---

166   Cicero, *De Oratore*
167   Seneca, *Letters From a Stoic*, Letter 2

Marcus Aurelius, *Meditations*[168]

"When a youth was giving himself airs in the Theatre and saying, 'I am wise, for I have conversed with many wise men,' Epictetus replied, 'I too have conversed with many rich men, yet I am not rich!'."

Epictetus, *Fragments*[169]

"Take care of this moment. Immerse yourself in its particulars. Respond to this person, this challenge, this deed. Quit the evasions. Stop giving yourself needless trouble. It is time to really live; to fully inhabit the situation you happen to be in now. You are not some disinterested bystander. Participate. Exert yourself."

Epictetus, *Discourses*

"The perfection of moral character consists in this, in passing every day as if it were the last, and in being neither violently excited nor torpid nor playing the hypocrite."

---

168 Marcus Aurelius, *Meditations*, IV
169 Epictetus, Fragment

Marcus Aurelius, *Meditations*

"When you do anything from a clear judgment that it ought to be done, never shrink from being seen to do it, even though the world should misunderstand it; for if you are not acting rightly, shun the action itself; if you are, why fear those who wrongly censure you?"

Enchiridion, *Discourses*[170]

"A man should always have these two rules in readiness. First, to do only what the reason of your ruling and legislating faculties suggest for the service of man. Second, to change your opinion whenever anyone at hand sets you right and unsettles you in an opinion, but this change of opinion should come only because you are persuaded that something is just or to the public advantage, not because it appears pleasant or increases your reputation."

Marcus Aurelius, *Meditations*[171]

"The condition and characteristic of an uninstructed person is this: he never expects from himself profit (advantage) nor harm, but from externals. The condition and characteristic of a philosopher is this: he expects all advantage and all harm from himself."

---

170   Enchiridion, *Discourses*, II, ibid
171   Marcus Aurelius, *Meditations*, III

Epictetus, *Enchiridion*

"Take care not to hurt the ruling faculty of your mind. If you were to guard against this in every action, you should enter upon those actions more safely."

Epictetus, *Enchiridion*[172]

"What can be more delightful than to have some one to whom you can say everything with the same absolute confidence as to yourself? Is not prosperity robbed of half its value if you have no one to share your joy?"

Cicero, *On Old Age*[173]

"For it is dangerous to attach one's self to the crowd in front, and so long as each one of us is more willing to trust another than to judge for himself, we never show any judgement in the matter of living, but always a blind trust, and a mistake that has been passed on from hand to hand finally involves us and works our destruction. It is the example of other people that is our undoing; let us merely separate ourselves from the crowd, and we shall be made whole. But as it is, the populace,, defending its own iniquity, pits itself against reason. And so

---

172   Epictetus, *Enchiridion*, 12, 22
173   Cicero, *On Old Age*

we see the same thing happening that happens at the elections, where, when the fickle breeze of popular favour has shifted, the very same persons who chose the praetors wonder that those praetors were chosen."

Seneca, *Letters From a Stoic*[174]

"It is one thing to put bread and wine away in a store-room, and quite another to eat them. What is eaten is digested and distributed around the body, to become sinews, flesh, bones, blood, a good complexion, sound breathing. What is stored away is ready at hand, to be sure, to be taken out and displayed whenever you wish, but you derive no benefit from it, except that of having the reputation of possessing it."

Epictetus, *Discourses*[175]

---

174   Seneca, *Letters From a Stoic*, Letter 23
175   Epictetus, *Discourses*, II

IV. Fall, A Time of Change

a. **The Stoics and Loss** – Stoics on dealing with life changes.

"Never let the future disturb you. You will meet it, if you have to, with the same weapons of reason which today arm you against the present."

Marcus Aurelius, *Meditations*

"Until we have begun to go without them, we fail to realize how unnecessary many things are. We've been using them not because we needed them but because we had them."

Seneca, *Letters From a Stoic*[176]

"Human life. Duration: momentary. Nature: changeable. Perception: dim. Condition of Body: decaying. Soul: spinning around. Fortune: unpredictable. Lasting Fame: uncertain. Sum Up: The body and its parts are a river, the soul a dream and mist, life is warfare and a journey far from home, lasting reputation is oblivion."

---

176   Seneca, *Letters From a Stoic*, Letter 22, Letter 19

Marcus Aurelius, *Meditations*

"Some people will say that memory fades away as the years pass. Of course it does if you don't exercise it or aren't very bright to begin with."

Cicero, *On Old Age*[177]

The gods are not to blame. They do nothing wrong, on purpose or by accident. Nor men either; they don't do it on purpose. No one is to blame.

Marcus Aurelius, *Meditations*[178]

"Remember that all we have is "on loan" from Fortune, which can reclaim it without our permission—indeed, without even advance notice. Thus, we should love all our dear ones, but always with the thought that we have no promise that we may keep them forever—nay, no promise even that we may keep them for long."

---

177   Cicero, *On Old Age*
178   Marcus Aurelius, *Meditations,* V, IV, I, ibid

Seneca, *Letters From a Stoic*

"And here are two of the most immediately useful thoughts you will dip into. First that things cannot touch the mind: they are external and inert; anxieties can only come from your internal judgment. Second, hat all these things you see will change almost as you look at them, and then will be no more. Constantly bring to mind all that you yourself have already seen changed. The universe is change: life is judgment."

Marcus Aurelius, *Meditations*

"All the greatest blessings are a source of anxiety, and at no time should fortune be less trusted than when it is best; to maintain prosperity there is need of other prosperity, and in behalf of the prayers that have turned out well we must make still other prayers. For everything that comes to us from chance is unstable, and the higher it rises, the more liable it is to fall. Moreover, what is doomed to perish brings pleasure to no one; very wretched, therefore, and not merely short, must the life of those be who work hard to gain what they must work harder to keep. By great toil they attain what they wish, and with anxiety hold what they have attained; meanwhile they take no account of time that will never more return."

Seneca, *Letters From a Stoic*

Time is a sort of river of passing events, and strong is its current; no sooner is a thing brought to sight than it is swept by and another takes its place, and this too will be swept away.

Marcus Aurelius, *Meditations*

"Remember two things: i. that everything has always been the same, and keeps recurring, and it makes no difference whether you see the same things recur in a hundred years or two hundred, or in an infinite period; ii. that the longest-lived and those who will die soonest lose the same thing. The present is all that they can give up, since that is all you have, and what you do not have you cannot lose."

Marcus Aurelius, *Meditations*

"And what's so bad about your being deprived of that?... All things seem unbearable to people who have become spoiled, who have become soft through a life of luxury, ailing more in the mind than they ever are in the body."

Seneca, *Letters From a Stoic*[179]

Remember that you ought to behave in life as you would at a banquet. As something is being passed around it comes to you; stretch out your hand, take a portion of it politely. It passes on; do not detain it. Or it has not come to you yet; do not project your desire to meet it, but wait until it comes in front of you. So act toward children, so toward a wife, so toward office, so toward wealth."

Epictetus, *Enchiridion*[180]

"In the life of a man, his time is but a moment, his being an incessant flux, his sense a dim rushlight, his body a prey of worms, his soul an unquiet eddy, his fortune dark, his fame doubtful. In short, all that is body is as coursing waters, all that is of the soul as dreams and vapors."

Marcus Aurelius, *Meditations*[181]

"That ever then the poor body of Socrates should have been dragged away and haled by main force to prison! That ever hemlock should have been given to the body of Socrates; that that should have breathed its life away!—Do you marvel at this? Do you hold this unjust? Is it for this that you accuse God? Had Socrates no

---

179 Seneca, *Letters From a Stoic*, Letter 13, Letter 9
180 Epictetus, *Enchiridion*, 18
181 Marcus Aurelius, *Meditations*, X, V, IV

compensation for this? Where then for him was the ideal Good? Whom shall we hearken to, you or him? And what says he?

"Anytus and Melitus may put me to death: to injure me is beyond their power."

And again:—

"If such be the will of God, so let it be."

Epictetus, *Discourses*[182]

"No one could endure lasting adversity if it continued to have the same force as when it first hit us. We are all tied to Fortune, some by a loose and golden chain, and others by a tight one of baser metal: but what does it matter? We are all held in the same captivity, and those who have bound others are themselves in bonds - unless you think perhaps that the left-hand chain is lighter. One man is bound by high office, another by wealth; good birth weighs down some, and a humble origin others; some bow under the rule of other men and some under their own; some are restricted to one place by exile, others by priesthoods: all life is a servitude.

---

182   Epictetus, *Discourses*

So you have to get used to your circumstances, complain about them as little as possible, and grasp whatever advantage they have to offer: no condition is so bitter that a stable mind cannot find some consolation in it."

Seneca, *Letters From a Stoic*[183]

"Is any man afraid of change? What can take place without change? What then is more pleasing or more suitable to the universal nature? And can you take a hot bath unless the wood for the fire undergoes a change? And can you be nourished unless the food undergoes a change? And can anything else that is useful be accomplished without change? Do you not see then that for yourself also to change is just the same, and equally necessary for the universal nature?"

Marcus Aurelius, *Meditations*

"Just as apples when unripe are torn from trees, but when ripe and mellow drop down, so it is violence that takes life from young men, ripeness from old. This ripeness is so delightful to me that, as I approach nearer to death, I seem, as it were, to be sighting land, and to be coming to port at last after a long voyage."

Cicero, *On Old Age*

---

183   Seneca, *Letters From a Stoic*

"How unlucky I am that this should happen to me. But not at all. Perhaps, say how lucky I am that I am not broken by what has happened, and I am not afraid of what is about to happen. For the same blow might have stricken anyone, but not many would have absorbed it without capitulation and complaint."

Marcus Aurelius, *Meditations*

"Everything is only for a day, both that which remembers and that which is remembered. "Observe constantly that all things take place by change, and accustom thyself to consider that the nature of the universe loves nothing so much as to change things which are and to make new things like them. For everything that exists is in a manner the seed of that which will be."

Marcus Aurelius, *Meditations*

"As I give thought to the matter, I find four causes for the apparent misery of old age; first it withdraws us from active accomplishments; second, it renders the body less powerful; third, it deprives us of almost all forms of enjoyment; fourth, it stands not far from death."

Cicero, *On Old Age*[184]

b. **Being Prepared** – Passages on reflecting on and preparing for shifts in fortune.

"Every part of me then will be reduced by change into some part of the universe, and that again will change into another part of the universe, and so on forever."

Marcus Aurelius, *Meditations*

"Perhaps the desire of the thing called fame torments you. See how soon everything is forgotten, and look at the chaos of infinite time on each side of the present, and the emptiness of applause, and the fickleness and lack of judgment in those who pretend to give praise, and the narrowness of its domain, and be quiet at last."

Marcus Aurelius, *Meditations*[185]

"The man who looks for the morrow without worrying over it knows a peaceful independence and a happiness beyond all others. Whoever has said, 'I have lived' receives a windfall every day he gets up in the morning."

---

184   Cicero, *On Old Age*
185   Marcus Aurelius, *Meditations,* II, I, ibid, IV, XII

Seneca, *Letters From a Stoic*[186]

"Treat what you don't have as nonexistent. Look at what you have, the things you value most, and think of how much you'd crave them if you didn't have them. But be careful. Don't feel such satisfaction that you start to overvalue them – that it would upset you to lose them."

Marcus Aurelius, *Meditations*

"If you want to make progress, put up with being perceived as ignorant or naive in worldly matters, don't aspire to a reputation for sagacity. If you do impress others as somebody, don't altogether believe it. You have to realize, it isn't easy to keep your will in agreement with nature, as well as externals. Caring about the one inevitably means you are going to shortchange the other."

Epictetus, *Discourses*

"Constantly recall those who have complained greatly about anything, those who have been most conspicuous by the greatest fame or misfortunes or enmities or fortunes of any kind: then think, where are they all now? Smoke and ash and a tale, or not even a tale."

---

186   Seneca, *Letters From a Stoic*, Letter 39

Marcus Aurelius, *Meditations*

"Luck is what happens when preparation meets opportunity."

Seneca, *Letters From a Stoic*

"Keep in mind how fast things pass by and are gone – those that are now and those to come. Existence flows past us like a river: the 'what' is in constant flux, the 'why' has a thousand variations. Nothing is stable, not even what's right here. The infinity of past and future gapes before us – a chasm whose depths we cannot see."

Marcus Aurelius, *Meditations*

"Here is your great soul—the man who has given himself over to Fate; on the other hand, that man is a weakling and a degenerate who struggles and maligns the order of the universe and would rather reform the gods than reform himself."

Seneca, *Letters From a Stoic*[187]

"Think of yourself as dead. You have lived your life. Now, take what's left and live it properly. What doesn't transmit light creates its own darkness."

---

[187] Seneca, *Letters From a Stoic*, Letter 10, Letter 18

Marcus Aurelius, *Meditations*[188]

"None of these things are foretold to me; but either to my paltry body, or property, or reputation, or children, or wife. But to me all omens are lucky, if I will. For whichever of these things happens, it is in my control to derive advantage from it."

Epictetus, *Discourses*[189]

"What would Heracles have been if he had said, "How am I to prevent a big lion from appearing, or a big boar, or brutal men?" What care you, I say? If a big boar appears, you will have a greater struggle to engage in; if evil men appear, you will free the world from evil men."

Epictetus, *Discourses*[190]

"The supreme ideal does not call for any external aids. It is homegrown, wholly self-developed. Once it starts looking outside itself for any part of itself it is on the way to being dominated by fortune."

Seneca, *Letters From a Stoic*

"Thou sayest, Men cannot admire the sharpness of thy wits.- Be it so: but there are many other things of which thou canst not say, I am not

---

188 Marcus Aurelius, *Meditations*, X, II, VI
189 Epictetus, *Discourses*, II, ibid
190 Epictetus, *Discourses*, II

formed for them by nature. Show those qualities then which are altogether in thy power, sincerity, gravity, endurance of labour, aversion to pleasure, contentment with thy portion and with few things, benevolence, frankness, no love of superfluity, freedom from trifling magnanimity. Dost thou not see how many qualities thou art immediately able to exhibit, in which there is no excuse of natural incapacity and unfitness, and yet thou still remainest voluntarily below the mark? Or art thou compelled through being defectively furnished by nature to murmur, and to be stingy, and to flatter, and to find fault with thy poor body, and to try to please men, and to make great display, and to be so restless in thy mind? No, by the gods: but thou mightest have been delivered from these things long ago. Only if in truth thou canst be charged with being rather slow and dull of comprehension, thou must exert thyself about this also, not neglecting it nor yet taking pleasure in thy dullness."

Marcus Aurelius, *Meditations*[191]

"Barley porridge, or a crust of barley bread, and water do not make a very cheerful diet, but nothing gives one keener pleasure than having the ability to derive pleasure even from that-- and the feeling of having arrived at something which one cannot be deprived of by any unjust stroke of fortune."

---

191   Marcus Aurelius, *Meditations*, X

Seneca, *Letters From a Stoic*[192]

"Seek not for events to happen as you wish but rather wish for events to happen as they do and your life will go smoothly."

Epictetus, *Enchiridion*

"For the only safe harbor in this life's tossing, troubled sea is to refuse to be bothered about what the future will bring and to stand ready and confident, squaring the breast to take without skulking or flinching whatever fortune hurls at us."

Seneca, *Letters From a Stoic*

"Your days are numbered. Use them to throw open the windows of your soul to the sun. If you do not, the sun will soon set, and you with it."

Marcus Aurelius, *Meditations*

"The greatest obstacle to living is expectancy, which hangs upon tomorrow and loses today. You are arranging what is in Fortune's control and abandoning what lies in yours."

---

192   Seneca, *Letters From a Stoic*, Letter 2, ibid, Letter 60

Seneca, *Letters From a Stoic*

"So you wish to conquer in the Olympic Games, my friend? And I, too... But first mark the conditions and the consequences. You will have to put yourself under discipline; to eat by rule, to avoid cakes and sweetmeats; to take exercise at the appointed hour whether you like it or not, in cold and heat; to abstain from cold drinks and wine at your will. Then, in the conflict itself you are likely enough to dislocate your wrist or twist your ankle, to swallow a great deal of dust, to be severely thrashed, and after all of these things, to be defeated."

Epictetus, *Discourses*[193]

"I was once a fortunate man but at some point fortune abandoned me.

But true good fortune is what you make for yourself. Good fortune: good character, good intentions, and good actions."

Marcus Aurelius, *Meditations*[194]

"Neither should a ship rely on one small anchor, nor should life rest on a single hope."

---

193   Epictetus, *Discourses*, II
194   Marcus Aurelius, *Meditations*, I, ibid

Epictetus, *Enchiridion*[195]

"Count your years and you'll be ashamed to be wanting and working for exactly the same things as you wanted when you were a boy. Of this one thing make sure against your dying day - that your faults die before you do. Have done with those unsettled pleasures, which cost one dear - they do one harm after they're past and gone, not merely when they're in prospect. Even when they're over, pleasures of a depraved nature are apt to carry feelings of dissatisfaction, in the same way as a criminal's anxiety doesn't end with the commission of the crime, even if it's undetected at the time. Such pleasures are insubstantial and unreliable; even if they don't do one any harm, they're fleeting in character. Look around for some enduring good instead. And nothing answers this description except what the spirit discovers for itself within itself. A good character is the only guarantee of everlasting, carefree happiness. Even if some obstacle to this comes on the scene, its appearance is only to be compared to that of clouds which drift in front of the sun without ever defeating its light."

Seneca, *Letters From a Stoic*[196]

"Another thing which will help you is to turn your mind to other thoughts and that way get away from your suffering. Call to mind

---

195 Epictetus, *Enchiridion*, 13
196 Seneca, *Letters From a Stoic*, Letter 16, Letter 31

things which you have done that have been upright or courageous; run over in your mind the finest parts you have played."

Marcus Aurelius, *Meditations*

"Never let the future disturb you. You will meet it, if you have to, with the same weapons of reason which today arm you against the present."

Marcus Aurelius, *Meditations*

"For I am not Eternity, but a human being—a part of the whole, as an hour is part of the day. I must come like the hour, and like the hour must pass!"

Epictetus, *Discourses* [197]

"It is always our choice whether or not we wish to pay the price for life's rewards. And often it is best for us not to pay the price, for the price might be our integrity."

---

197   Epictetus, *Discourses*, III

Epictetus, *Fragments*

"Look back over the past, with its changing empires that rose and fell, and you can foresee the future too."

Marcus Aurelius, *Meditations*[198]

"Each day acquire something that will fortify you against poverty, against death, indeed against other misfortunes as well; and after you have run over many thoughts, select one to be thoroughly digested that day."

Seneca, *Letters From a Stoic*[199]

c. **Dealing with Death** – The Stoics and death, personal and impersonal.

"It is more necessary for the soul to be cured than the body; for it is better to die than to live badly."

---

198   Marcus Aurelius, *Meditations*, II, IV, II
199   Seneca, *Letters From a Stoic*, Letter 12

Epictetus, *Fragments*[200]

"The gods either have power or they have not. If they have not, why pray to them? If they have, then instead of praying to be granted or spared such-and-such a thing, why not rather pray to be delivered from dreading it, or lusting for it, or grieving over it? Clearly, if they can help a man at all, they can help him in this way. You will say, perhaps, 'But all that is something they have put in my own power.' Then surely it were better to use your power and be a free man, than to hanker like a slave and a beggar for something that is not in your power. Besides, who told you the gods never lend their aid even towards things that do lie in our own power? Begin praying in this way, and you will see. Where another man prays 'Grant that I may possess this woman,' let your own prayer be, 'Grant that I may not lust to possess her.' Where he prays, 'Grant me to be rid of such-and-such a one,' you pray, 'Take from me my desire to be rid of him.' Where he begs, 'Spare me the loss of my precious child,' beg rather to be delivered from the terror of losing him. In short, give your petitions a turn in this direction, and see what comes."

---

[200] Epictetus, Fragment

Marcus Aurelius, *Meditations*

"It is our attitude toward events, not events themselves, which we can control. Nothing is by its own nature calamitous -- even death is terrible only if we fear it."

Epictetus, *Enchiridion*

"Finally, waiting for death with a cheerful mind, as being nothing else than a dissolution of the elements of which every living being is compounded. But if there is no harm to the elements themselves in each continually changing into another, why should a man have any apprehension about the change and dissolution of all the elements?"

Marcus Aurelius, *Meditations*[201]

"Men are disturbed not by things, but by the views which they take of things. Thus death is nothing terrible, else it would have appeared so to Socrates. But the terror consists in our notion of death, that it is terrible. When, therefore, we are hindered, or disturbed, or grieved let us never impute it to others, but to ourselves; that is, to our own views. It is the action of an uninstructed person to reproach others for his own misfortunes; of one entering upon instruction, to reproach

---

201   Marcus Aurelius, *Meditations*, IX, II, VII

himself; and of one perfectly instructed, to reproach neither others or himself."

Epictetus, Enchiridion[202]

"For it is not death or pain that is to be feared, but the fear of pain or death."

Epictetus, *Fragments*

"That which has died falls not out of the universe. If it stays here, it also changes here, and is dissolved into its proper parts, which are elements of the universe and of thyself. And these too change, and they murmur not"."

Marcus Aurelius, *Meditations*

"Death is a release from the impressions of the senses, and from desires that make us their puppets, and from the vagaries of the mind, and from the hard service of the flesh."

---

202   Epictetus, *Enchiridion*, 10

Marcus Aurelius, *Meditations*

"What man can you show me who places any value on his time, who reckons the worth of each day, who understands that he is dying daily? For we are mistaken when we look forward to death; the major portion of death has already passed. Whatever years be behind us are in death's hands."

Seneca, *Letters From a Stoic*[203]

"[Do not get too attached to life] for it is like a sailor's leave on the shore and at any time, the captain may sound the horn, calling you back to eternal darkness."

Epictetus, *Discourses*[204]

"A certain Spartan, whose name hasn't even been passed down, despised death so greatly that when he was being led to execution after his condemnation by the ephors, he maintained a relaxed and joyous expression. To an enemy's challenge – 'Is this how you mock the laws of Lycurgus?' – he answered, 'On the contrary, I give great thanks to him, for he decreed a punishment that I can pay without taking out a loan or juggling debts.' O worthy man of Sparta! His spirit was so great

---

203   Seneca, *Letters From a Stoic*, Letter 51
204   Epictetus, *Discourses*, III

that it seems he must have been an innocent man condemned to die. There have been many such in our own country."

Cicero, *On Living and Dying Well*[205]

"Wherefore it is a shame for man to begin and to leave off where the brutes do. Rather he should begin there, and leave off where Nature leaves off in us: and that is at contemplation, and understanding, and a manner of life that is in harmony with herself. See then that you do not die without being spectators of these things."

Epictetus, *Enchiridion*[206]

"Just that you do the right thing. The rest doesn't matter. Cold or warm. Tired or well-rested. Despised or honored. Dying...or busy with other assignments. Because dying, too, is one of our assignments in life. There as well: "To do what needs doing." Look inward. Don't let the true nature of anything elude you. Before long, all existing things will be transformed, to rise like smoke (assuming all things become one), or be dispersed in fragments...to move from one unselfish act to another with God in mind. Only there, delight and stillness...when jarred, unavoidably, by circumstances, revert at once to yourself, and

---

205   Cicero, *On Living and Dying Well*
206   Epictetus, *Enchiridion*, 24

don't lose the rhythm more than you can help. You'll have a better grasp of the harmony if you keep going back to it."

Marcus Aurelius, *Meditations*[207]

"If you shall be afraid not because you must some time cease to live, but if you shall fear never to have begun to live according to nature – then you will be a man worthy of the universe that has produced you, and you will cease to be a stranger in your native land."

Marcus Aurelius, *Meditations*

"It is possible to learn the will of nature from the things in which we do not differ from each other. For example, when someone else's little slave boy breaks his cup we are ready to say, "It's one of those things that just happen." Certainly, then, when your own cup is broken you should be just the way you were when the other person's was broken. Transfer the same idea to larger matters. Someone else's child is dead, or his wife. There is no one would not say, "It's the lot of a human being." But when one's own dies, immediately it is, "Alas! Poor me!" But we should have remembered how we feel when we hear of the same thing about others."

---

207   Marcus Aurelius, *Meditations*, X, ibid

Epictetus, *Enchiridion*[208]

"Do not act as if you were going to live ten thousand years. Death hangs over you. While you live, while it is in your power, be good."

Marcus Aurelius, *Meditations*

"Did not he, then, who, if he had died at that time, would have died in all his glory, owe all the great and terrible misfortunes into which he subsequently fell to the prolongation of his life at that time?"

Cicero, *On Old Age*[209]

"People are delighted to accept pensions and gratuities, for which they hire out their labour or their support or their services. But nobody works out the value of time: men use it lavishly as if it cost nothing. But if death threatens these same people, you will see them praying to their doctors; if they are in fear of capital punishment, you will see them prepared to spend their all to stay alive."

---

208   Epictetus, *Enchiridion*
209   Cicero, *On Old Age*

Seneca, *On The Shortness of Life*[210]

"Brief is man's life and small the nook of the earth where he lives; brief, too, is the longest posthumous fame, buoyed only by a succession of poor human beings who will very soon die and who know little of themselves, much less of someone who died long ago."

Marcus Aurelius, *Meditations*[211]

"What is death? A "tragic mask." Turn it and examine it. See, it does not bite. The poor body must be separated from the spirit either now or later, as it was separated from it before. Why, then, are you troubled, if it be separated now? for if it is not separated now, it will be separated afterward. Why? That the period of the universe may be completed, for it has need of the present, and of the future, and of the past. What is pain? A mask. Turn it and examine it. The poor flesh is moved roughly, then, on the contrary, smoothly. If this does not satisfy you, the door is open: if it does, bear. For the door ought to be open for all occasions; and so we have no trouble."

Epictetus, *Discourses*[212]

---

210   Seneca, *On The Shortness of Life*
211   Marcus Aurelius, *Meditations*, II, IV
212   Epictetus, *Discourses*, III

"Don't let yourself forget how many doctors have died, furrowing their brows over how many deathbeds. How many astrologers, after pompous forecasts about others' ends. How many philosophers, after endless disquisitions on death and immortality. How many warriors, after inflicting thousands of casualties themselves. How many tyrants, after abusing the power of life and death atrociously, as if they were themselves immortal.

How many whole cities have met their end: Helike, Pompeii, Herculaneum, and countless others.

And all the ones you know yourself, one after another. One who laid out another for burial, and was buried himself, and then the man who buried him - all in the same short space of time.

In short, know this: Human lives are brief and trivial. Yesterday a blob of semen; tomorrow embalming fluid, ash.

To pass through this brief life as nature demands. To give it up without complaint.

Like an olive that ripens and falls.

Praising its mother, thanking the tree it grew on."

Marcus Aurelius, *Meditations*[213]

"It is not that we have so little time but that we lose so much. ... The life we receive is not short but we make it so; we are not ill provided but use what we have wastefully."

Seneca, *On The Shortness of Life*[214]

"The history of your life is now complete and your service is ended: and how many beautiful things you have seen; and how many pleasures and pains you have despised; and how many things called honorable you have spurned; and to how many ill-minded folks you have shown a kind disposition."

Marcus Aurelius, *Meditations*[215]

---

213   Marcus Aurelius, *Meditations*
214   Seneca, *On The Shortness of Life*
215   Marcus Aurelius, *Meditations*, IV, V

BIBLIOGRAPHY

Epictetus. *The Enchiridion*. Translated by Elizabeth Carter. http://classics.mit.edu/Epictetus/epicench.html.

Aurelius, Marcus. *Meditations*. Translated by George Long. http://classics.mit.edu/Antoninus/meditations.html.

Cicero, Marcus. *On Old Age*. Translated by Robert Allison, 1916. *Internet Archive*, archive.org/stream/cu31924026475230/cu31924026475230_djvu.txt.

Cicero, Marcus T. *On Duties*. Translated by Cyrus R. Edmonds. New York, NY: Harper & Brothers, 1855. https://archive.org/stream/cicerosthreebook00cicerich/cicerosthreebook00cicerich_djvu.txt.

Epictetus. *Discourses*.

http://classics.mit.edu/Epictetus/discourses.html.

Seneca, Lucius. *On the Shortness of Life: Life Is Long If You Know How to Use It (Penguin Great Ideas)*. Translated by C.D.N. Costa, 1st ed., Penguin Classics.

Seneca, Lucius. *Letters from a Stoic (Penguin Classics)*. Translated by Robin Campbell, Penguin Classics, 1969.

www.ingramcontent.com/pod-product-compliance
Lightning Source LLC
Chambersburg PA
CBHW070047230426
43661CB00005B/792